SOTHIS PRESS

Sothis Press aims to completely reimagine the public domain book with an emphasis on universal access to captivatingly designed print editions. We believe that technology will enable every single book ever published to be perpetually accessible in beautiful print form. Sothis is beauty in ubiquity.

www.sothis.press
1 Waterhouse Square, London, ECIN 2ST

The Central Railroad of New Jersey

An Illustrated Guide-Book (With Road-Maps)

Gustav Kobbé

SOTHIS PRESS

ISBN 978-1-0365-0134-1

The Central Railroad of New Jersey
is in the public domain.

This edition is published by
Sothis Press © 2023. The company
is registered in England and Wales.
The Company Number is 14900389.

LAKE HOPATCONG

THE
CENTRAL RAILROAD
OF
NEW JERSEY.

AN ILLUSTRATED GUIDE-BOOK (WITH ROAD-MAPS).

BY

GUSTAV KOBBÉ.

1890.
GUSTAV KOBBÉ.
251 Broadway, New York.

PREFACE.

This aims to be an accurate descriptive guide-book to those parts of Central and Northern New Jersey that are reached by the main stem of the Central Railroad of New Jersey and its branches.

Historical matter relating to these sections has been carefully collected, and the most striking incidents have been incorporated in the work in order that the book may have romantic as well as descriptive and statistical interest.

I shall esteem it a favor if any one who may discover any errors of commission or omission will call my attention to them.

GUSTAV KOBBÉ.

SHORT HILLS,
ESSEX CO., N. J.

The illustrations are from drawings by Marie Olga Kobbé and F. A. Feraud; from aquarelles by Hugh Smythe; direct from photographs; and from views of Lake Hopatcong kindly furnished by the Hotel Breslin.

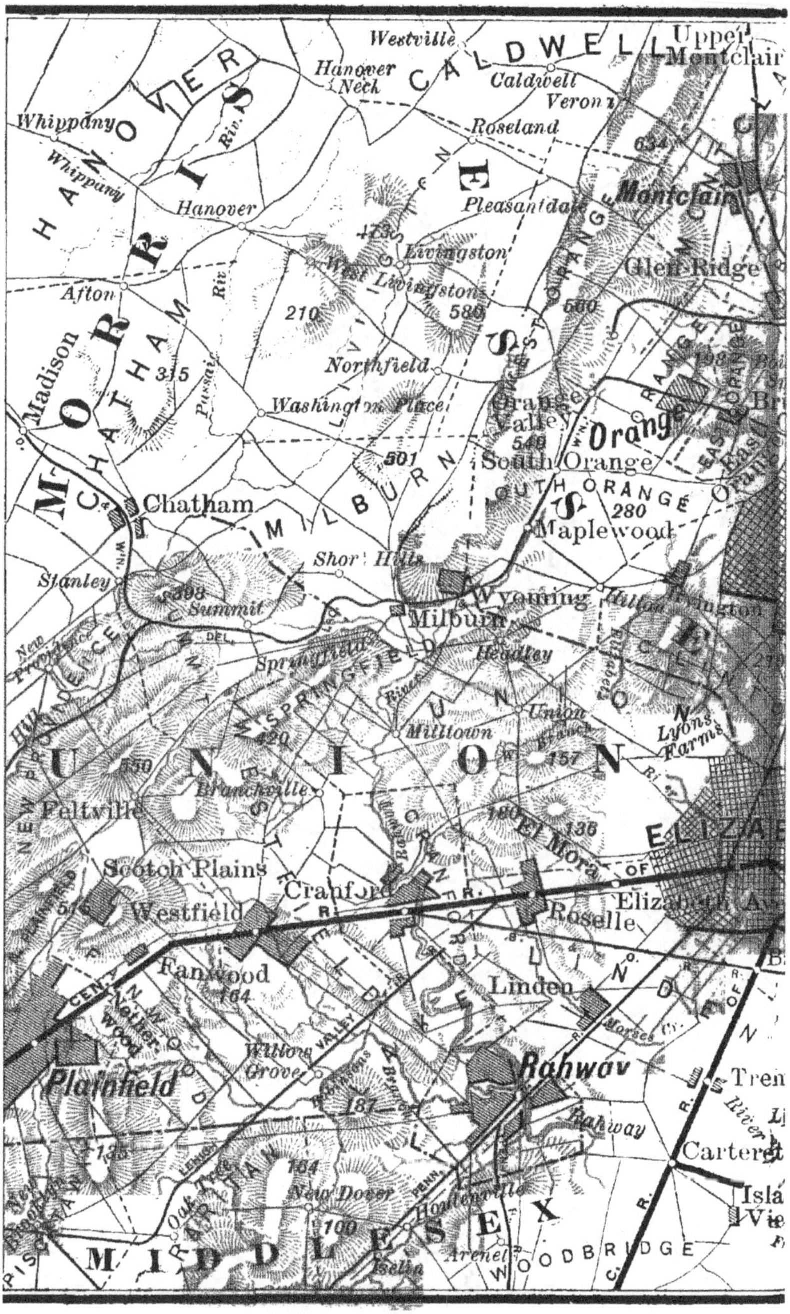

Westville
CALDWELL
Upper Montclair
Hanover Neck
Caldwell
Whippany
HANOVER
Whippany Riv.
Roseland
634
Hanover
Pleasantdale
Montclair
LIVINGSTON
Livingston
West Livingston
Glen Ridge
Afton
210
580
560
Madison
CHATHAM
315
Passaic Riv.
Northfield
Washington Place
Orange Valley
Orange
540
South Orange
501
MILBURN
SOUTH ORANGE
280
Chatham
Maplewood
Short Hills
Stanley
398
Summit
Wyoming
Irvington
Milburn
Headley
New Providence
Springfield
SPRINGFIELD
Union
Milltown
Lyons Farms
420
550
157
UNION
Branchville
Peltville
180
136
El Mora
Scotch Plains
Cranford
Elizabeth Av.
Westfield
Roselle
WESTFIELD
Fanwood
164
Linden
LINDEN
Netherwood
Willow Grove
Rahway
Morses Cr.
Plainfield
187
Rahway
Carteret
135
164
New Dover
Houtenville
MIDDLESEX
100
Oak Tree
Arenel
Iselin
WOODBRIDGE
MORRIS
ESSEX

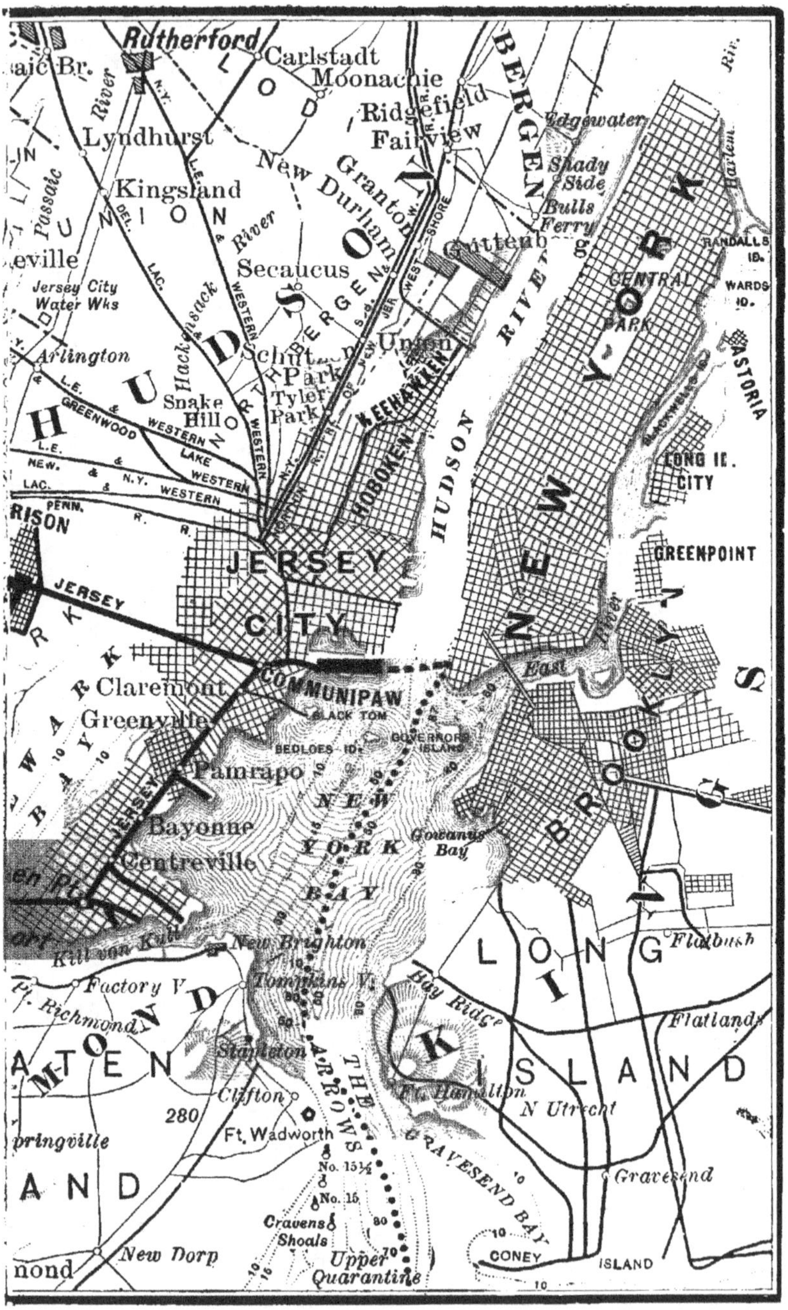
Rutherford
Carlstadt
Moonachie
Ridgefield
Fairview
BERGEN
Edgewater
Shady Side
Bulls Ferry
Lyndhurst
Kingsland
New Durham
Granton
Passaic
River
Hackensack River
Secaucus
Jersey City Water Wks
Arlington
Snake Hill
Tyler Park
Union
HUDSON
NORTH BERGEN
HOBOKEN
WEEHAWKEN
HUDSON RIVER
NEW YORK
CENTRAL PARK
RANDALLS ID.
WARDS ID.
ASTORIA
LONG ID. CITY
GREENPOINT
JERSEY CITY
COMMUNIPAW
BLACK TOM
BEDLOES ID.
GOVERNORS ISLAND
Claremont
Greenville
NEWARK BAY
Pamrapo
Bayonne
Centreville
NEW YORK BAY
Gowanus Bay
East River
BROOKLYN
Kill von Kull
New Brighton
Tompkins V.
Factory V.
Pt. Richmond
STATEN ISLAND
Stapleton
Clifton
Ft. Wadworth
THE NARROWS
Bay Ridge
Ft. Hamilton
KINGS
LONG ISLAND
Flatbush
Flatlands
N Utrecht
Gravesend
GRAVESEND BAY
CONEY ISLAND
No. 15½
No. 15
Cravens Shoals
Upper Quarantine
New Dorp
Springville
280
L.E. & WESTERN
GREENWOOD LAKE
NEW & N.Y. LAC. WESTERN
PENN. R. R.
JERSEY
WEST SHORE

INTRODUCTION.

[The Author will esteem it a favor if his attention is called to any errors of omission or commission.]

Topography and Geology.—New Jersey lies on the eastern slope of the Appalachian region. The eastern base of this system is a plain sloping gently toward the Atlantic, with an elevation along the base of the mountains of from 200 to 600 feet in New Jersey. Those portions of the State described in this book lie chiefly within the limits of the Triassic Red Sandstone Plain and the Azoic Central Highland Plateau. Remarkable features of the Red Sandstone Plain are its prominent ridges of trap-rock.

Bergen Point is a ridge of trap, a continuation of the Palisades of the Hudson. The Newark and New York branch of the Central Railroad of New Jersey cuts through this ridge and, after crossing the Hackensack and Passaic and traversing the tidal meadows, enters the Red Sandstone Plain at Newark. The branch from Newark to Elizabeth runs for the entire distance after leaving the line of the Newark and New York branch to Elizabethport through the tidal meadows.

The main stem follows the trap-rock ridge almost the entire length of Bergen Point. Then, crossing Newark Bay, it enters at Elizabethport the Red Sandstone Plain, which it traverses to Lebanon, where it penetrates the southwestern end of the Highlands to Phillipsburg on the Delaware, running from Westfield to Somerville within full view of the Watchung Mountains of trap-rock. From the vicinity of Pompton these two ridges run southwest and continue for 40 miles exactly parallel, with their crests one and one-half miles apart. Each has a steep eastern slope and a long, gentle western one, with remarkably level crests. The eastern ridge is called First, the western Second Mountain, and they have various local names. Between Summit and Bound Brook the valley between the ridges pours its drainage through three narrow gaps in First Mountain—Green Brook at Scotch Plains; Stony Brook at Plainfield; and Middle Brook at Bound Brook. This valley is called Washington Valley. It is narrow at Summit, but widens back of Bound Brook at Martinsville. At Milburn there is a gap of two miles in First Mountain, which then rises to 546 feet, but southwest of this its crest keeps below 500 feet, though at Washington Rock, back of Plainfield, it attains a height of 539 feet. Second Mountain rises at Summit to 547 feet, and eighteen miles southwest to 653 feet.

At Martinsville both ridges turn a right angle and run back northwest toward the Highlands, which are only six miles beyond. First Mountain falls off at Pluckamin. Second Mountain continues north from this point almost to the Highlands, then curves back northeast. Just at this point is the only pass through Second Mountain, excepting the outlet of the valley, at Paterson, which is lower than the pass at Summit. Were this pass (Moggy Hollow) closed and a dam one and one-half miles long raised across the outlet at Totowa, near Paterson, a lake would be formed, between Second Mountain and the Highlands, of 300 square miles, 200 feet deep in the deeper parts and 385 feet above the sea. A series of distinct gravel terraces within this area, with elevations close to 400 feet, show that at some time in the glacial epoch such a lake actually existed.

Cushetunk Mountain near White House and Lebanon is also a trap formation. It forms a horseshoe, sweeping about an ellipsoidal valley (Round Valley) which is completely encircled by it and the gneiss hills at the west. Its maximum elevation is 767 feet. The length of the semi-circular sweep of trap is about seven miles.

The soil of the Red Sandstone Plain is remarkably fertile, and this section is the most densely populated of New Jersey. The formation contains few organic remains. Fossil fishes, footprints of what probably were air-breathing animals of the Reptilian age, and plants evidently of a higher order than those belonging to the Carboniferous age, have been discovered.

The Red Sandstone Plain is of sedimentary origin,

and its entire area is believed to have been as high as the tops of the trap ridges which now rise up so boldly out of the plain. But it has been worn away, probably by glacial erosion, to a depth of from 300 to 500 feet. The trap, which is of igneous origin, is supposed to have been sandwiched between two layers of sandstone, the upper of which gave the ridges a height nearly as great again as they now possess. When this upper layer had been worn away, the hardness of the trap defeated the power of the eroding agent, and the trap ridges remained while the rest of the plain was worn away to the depths mentioned above. The long line of Short Hills in Middlesex County which stretches between the two mountain gaps at Scotch Plains and Plainfield seems to be composed of the drift which the eroding agency forced through these notches. In the gap in First Mountain in Essex County is another locality similarly formed, and, oddly enough, similarly named.

The southern limit of the great sheet of ice which once covered New Jersey, in common with the rest of our continent, down to latitude 40° 30′, has been defined in the reports of the Geological Survey of New Jersey, and its terminal moraine clearly traced across the State. It is distinctly marked by a line of extremely irregular, fantastically-arranged hills of gravel and boulders formed of the material eroded by the glacier from the hills to the north and deposited here where the ice melted. This moraine begins at Perth Amboy and runs thence through Metuchen, east of Plainfield, where the Netherwood Hotel is built upon it, to the base of the First Mountain north of Scotch Plains. Thence the mantle of gravel is wrapped about the slope and over the north end of Springfield or Roll's Hill and filling completely the valley west, crosses Second Mountain and lies up against the north end of Long Hill, at Chatham. From here to Morristown it fills the valley of the Passaic with a broad ridge of gravel; thence it skirts around the base of the Highlands and up through the valley of the Rockaway to Dover. From here the line is quite direct by Budd's Lake, Hackettstown and Townsbury, to Belvidere.

To the north of this the ice sheet was thick enough to overtop all of the mountains of northern New Jersey and most of those of New York. Its movement was generally toward the south, and when we recall that a thickness of 2,000 feet would mean a pressure at the base of sixty tons per square foot, and that often boulders were imbedded in the base of the ice and moved forward with irresistible force, it may bring some conception of the enormous eroding action of the glacier. It denuded the ridges of all disintegrated rocks, scooping out transverse depressions where the rock was soft, and leaving often hard, bare summits and irregular, jagged ridge lines in place of the well-soiled, gracefully-undulating ridges to the south of the moraine. It deposited in the valleys great masses of gravel and mud, which have been in some cases assorted and worked down into level terraces by water, but again left in all the fantastic disorder of their original deposition, in crooked ridges enclosing bowl-like depressions with no outlets, or hills carrying similar depressions in their very tops, like small volcanoes with their craters, and in every conceivable topographically-monstrous arrangement. Often these deposits have closed the outlet of a valley, holding back the water in beautiful lakes and ponds, the water having been forced back over the original divide of the valley into another drainage system. When the drift dam has not been high enough for this, it has been cut away again by the water overtopping it. Remains of such dams may be found, with gravel terraces on the slopes of the valley above to mark the shores of the ancient lakes. The above accounts for the existence of most of the beautiful lakes of the northern counties, and also for the swamps and sink-holes which are merely shallow lake basins which have become filled with mud or vegetable matter. The drift dam which has formed Budd's lake is very evident, as is the one at Green Pond. The slopes of the hills of this region have usually been left covered with boulders, the finer material having been carried down into the valleys by water, and the whole aspect of the country has been changed.

The branch which the Central Railroad of New Jersey sends north from High Bridge penetrates into the very heart of the Highlands. Their surface varies from 175 feet to 1,200 feet above the sea level. The chief mountain-ranges in point of size are Schooley's and Green

Pond. Magnetic iron ore is found in abundance throughout this region. Of the deposits west of the Green Pond Mountain those of the Hurd and Ford Mines are the most important.

These ores occur in the Azoic rocks which underlie by far the greatest portion of this territory. They are not veins, but are of sedimentary origin. It is supposed that proto-salts of iron were leached out of the rocks by water and carried by it to some pond hole, there forming metallic films which would successively sink of their own weight, forming an ever-thickening layer of iron rust at the bottom of the pond hole. Afterward, through some action of nature, such as subsidence or elevation of the surrounding country, an influx of mud and sand covered the rust, forcing the greater part of the water off and solidifying the layer. Limonite in this form occurs at Beattystown. Further agencies converted this limonite into magnetic iron ore, and through convulsions of nature what once formed the bottoms of lakes may have become hills or mountains.

Another geological period represented by the rocks of the Highlands is the Lower Silurian, which includes the deposit of Potsdam sandstone near Flanders, where it occurs in the form of a fine white sand, which, being very refractory, is extensively used as a lining for furnaces. To this period also belongs the magnesian limestone, the largest deposit of which is found in German Valley. It is about nine miles long by one-half mile wide, extending from a mile northeast of Naughright to a mile southwest of Califon, and is much worked for lime for farming purposes and for use in the blast furnaces at Chester and Boonton. There is then a break in the geological succession, the next period represented being the Triassic, or new Red Sandstone.

This portion of the Highlands, which is known as the Central Highland Plateau, has a width of from five to seven miles from the New York line to Lake Hopatcong and Budd's Lake, but tapers down irregularly to a ridge two miles wide near the Delaware. The Central Railroad attains at the Ogden Mines the highest elevation reached by any railroad in the State—1,240 feet. Just north of Ford and Scofield Mines the plateau is 1,396 feet high, the maximum for this section of the State.

Lake Hopatcong, with a surface-elevation of 926 feet, lies right in the middle of the plateau.

At the south end of Lake Hopatcong is an important pass, at the same elevation as the surface of the lake; and it is worthy of note that, while the natural outlet of the lake is westward by the Musconetcong, the dam erected across this outlet by the Morris Canal Co. would have sent its water coursing eastward into the Raritan water-shed, had not a small side dam been raised at the extreme south end of the lake. Southwest of this pass is Schooley's, and beyond this Musconetcong Mountain. In popular parlance, however, Lake Hopatcong is located on Schooley's Mountain, the name being extended so as to cover the plateau northeast of the pass.

NATURAL HISTORY.—The Minerals of chief importance in the territory covered by this book are the Magnetic Iron Ores along the route of the Ogden Mine Railroad (a part of the High Bridge branch of the Central Railroad of New Jersey), at High Bridge, Chester and Port Oram; Limonite at Califon, Limestone at Vernoy, Granite at German Valley, and Copper Ores near Somerville.

The Fauna is that of the Middle States; but special mention may be made of the black bass, Oswego bass, pickerel and perch in some of the lakes, and the trout in some of the streams; and of the game birds. These are, however, mentioned in the body of the book. The typical mountain butterfly, *Limin Arthemis*, is found in the Highlands, and *Satiris Nephele* replaces *Satiris Alope* of the low lands.

The Flora of this region is in general similar to that of the country adjacent to New York. The bogs about Budd's Lake and Succasunna contain some plants that are not known elsewhere in the State; for instance, *Potentilla palustris, Salis myrtilloides, Rhododendron Canadense.* The aquatic flora of the larger bodies of water, Hopatcong and Green Pond, and, in a lesser degree, Budd's Lake, is extremely interesting and has been carefully explored by botanists. The higher mountain summits are inhabited by plants of a more northerly range, a great many interesting species being found there that do not occur on the low lands; among these being *Potentilla arguta, Clematis verticillaris, Coptis trifolia, Gillenia trifoliata* and *Genru rivale.*

The forests of the region are abundant, and much land that cannot be used for farming purposes is devoted to woodland. Birch, oak, maple and hickory are found in abundance and there are a few groves of white pine, which, however, is not so often seen as further south on the coast.

CLIMATE.—Summer in the Highlands is not marked by so great extremes of heat, and hence the weather is much more endurable than in the cities. The attractiveness of Schooley's Mountain, Lake Hopatcong, Budd's Lake, Chester and other Highland resorts is no doubt owing to the absence of excessively high temperature in midsummer; or more especially, perhaps, to the more markedly lower night temperature. On the Red Sandstone Plain, where the thermometric record may show but little difference between the temperature of the city and country, there is no doubt that the more open situation of the latter, which allows a free circulation of air, is an advantage which offsets mere temperature.

HISTORY.—August 16, 1609, Henry Hudson entered Delaware Bay. Finding the navigation difficult on account of shoal water, he changed his course, and, following the eastern shore of New Jersey, anchored his ship (the *Half Moon*) in Sandy Hook Bay September 3, 1609. September 11th he sailed through the Narrows and discovered the river which bears his name. October 4 he again set sail for Europe. Hudson was in the employ of the Dutch East India Company, which in 1610 despatched a vessel to the scene of his discoveries and established a fort and trading-house on Manhattan Island, which was called New Amsterdam.

It appears that as early as 1614 a redoubt was thrown up on the right bank of the Hudson, probably at the present Jersey City point; and, in 1618, Bergen is believed to have been settled by a number of Danes or Norwegians who accompanied the Dutch colonists to New Netherlands.

In 1664, Charles II of England resolved upon the reduction of New Netherlands, and despatched a fleet under Sir Robert Carr and Colonel Richard Nichols, to whom the Dutch Governor, Stuyvesant, was forced to surrender. Meanwhile King Charles had made an extensive grant of colonial territory to his brother, the

Duke of York, who in turn conveyed to Lord Berkeley and Sir George Carteret that portion of the grant which is now the State of New Jersey. This name being given to the new province in honor of Carteret, who had held the Island of Jersey for the King during his contest with Parliament.

These two Proprietors appointed Philip Carteret Governor of New Jersey. He came over in 1665 and made Elizabethtown the seat of government. Some disputes broke out between him and prior settlers, who, having purchased their land direct from the Indians, refused to recognize the right of the Proprietors to collect rent, the troubles culminating in 1672 in an insurrection. Carteret was obliged to seek redress in England, and, during his absence, his officers were imprisoned, their estates confiscated and James Carteret, a dissolute natural son of the Governor, prevailed upon to usurp the government.

July 30, 1673, war having broken out between England and Holland, a Dutch squadron appeared before New York, and, in the absence of Governor Lovelace, the place surrendered. But the following spring, by treaty of peace, New York and New Jersey came again under English dominion, and Major Edmund Andross was sent over as Governor of New York by the Duke of York. As the Governor sought to extend his authority also over New Jersey, the people, on Governor Carteret's return in 1675, made common cause with him against Andross.

Early in 1673, Lord Berkeley, having become dissatisfied with the financial returns from his venture in colonial lands, sold his interests in New Jersey to two Quakers, John Fenwick and Edwin Billynge, and the province was, in 1676, divided into East and West Jersey, the division line running from the east side of Little Egg Harbor straight north to the Delaware.

Sir George Carteret died in 1679, and soon afterwards East Jersey was sold to pay his debts to a "syndicate" of twelve. Philip Carteret continued Governor until about 1681. Difficulties between the Proprietors and those who claimed under direct purchase from the Indians were continuous, both in East and West Jersey, and in 1702 the Proprietors grew so weary of the incessant strife, that they surrendered their rights to the Crown.

Queen Anne immediately reunited the Jerseys, and appointed her kinsman, Edward Hyde, Lord Cornbury, Governor of both New York and New Jersey. The province was governed from New York until 1738, when a commission arrived to Lewis Morris as Governor of New Jersey separate from New York. The last of the Royal Governors was William Franklin, a natural son of Benjamin Franklin. Events in the Revolution which took place within the limits of the territory covered by this work are related in their proper places in the body of the book.

INDIAN HISTORY.—The aborigines whom the white settlers found in New Jersey were a portion of the Delaware Nation. They were so called by the whites, but were known among themselves as the Lenni Lenapè Nation, and were divided into Unamis or Turtles, Unalachtos or Turkeys, and Minsi or Wolfs—the fiercest of the tribes. These last dwelt in Northern New Jersey. These tribes were in turn subdivided into families, among them the Navesinks, Assanpinks, Matas, Shackamaxons, Chichequaas (Cheesequakes), Raritans, Nanticokes, Tutelos and Nariticongs.

There were two Indian paths from the interior to the coast which in the early days were used by the whites as highways—the Minnisink and Burlington paths. The former, starting at Minnisink, on the upper Delaware, passed through Sussex, Morris, Union and Middlesex Counties, crossed the Raritan by a ford about three miles above its mouth, and ran through the village of Middletown to Clay Pit Creek on the Navesink, and thence to the mouth of that river. The Burlington path started from Crosswicks, at a junction of two paths, respectively from Trenton and Burlington; ran to Freehold, whose main street is on the old path, and thence toward Middletown, near which place it joined the Minnisink path. A branch from below Freehold led through Tinton Falls to Long Branch. The only Indian settlements whose sites have been identified are that at Crosswicks, one not far from the Navesink ford on the Raritan, and one at Lake Hopatcong. The aborigines in the State of New Jersey did not at any time after the white discovery number over 2,000.

The government of the province always recognized the

title of the Indians to the lands, and always insisted on a fair purchase of lands from them. For this reason the white settlers never had trouble with the aborigines. In 1758, most of the Indians having sold their land agreed to the extinguishment of most of their titles, except the right to fish in all the rivers and bays south of the Raritan, and to hunt on all uninclosed lands. In 1802 they removed to New Stockbridge, near Oneida Lake, N. Y. In 1832 the remnant of the Lenni Lenapès, forty in number, were settled at Statesburgh, on Fox River, Wis. Believing that they had never parted with the right to fish and hunt secured to them in 1758, they deputed one of their number, Wilted Grass, known among the whites as Bartholomew S. Calvin, who had served with credit under Washington, to lay their claim before the New Jersey Legislature. This he did in a memorial, couched in language simple and pathetic, beginning: "I am old and weak and poor, and therefore a fit representative of my people. You are young and strong and rich, and therefore fit representatives of your people." The Legislature voted $2,000, the sum asked for. Wilted Grass addressed a letter of thanks to the Legislature in which the following noteworthy passage occurred:

"Not a drop of our blood have you spilled in battle; not an acre of our land have you taken but by our consent. These facts speak for themselves and need no comment. They place the character of New Jersey in bold relief and bright example to those States within whose territorial limits our brethren still remain. Nothing but benisons can fall upon her from the lips of a Lenni Lenapè."

TRANSPORTATION.—It may be said that thousands of the best citizens of New York are not citizens of that city at all. In the morning they flood the business districts of the metropolis; in the evening they ebb away. They are citizens of New York, in so far as the city owes to their brains and energy a great share of its prosperity; they are not citizens, in that they live and vote elsewhere. If this great suburban army of intelligent men lived as well as worked in New York, we would probably hear less of the necessity of municipal reform, for there would be just so much more intelligence among the voting popu-

JERSEY CITY TERMINUS

lation—which brings us back to our starting point: that of New York's best citizens thousands are, unfortunately for it, citizens of New Jersey, Long Island and other suburban districts.

Among the most intelligent and progressive of these non-citizen citizens are those who reach the city by the Central Railroad of New Jersey—doubtless because of the charming and healthful surroundings amid which they have chosen their homes, and also because they are conveyed to and from their places of business by a railroad which combines the greatest speed consistent with safety with the greatest comfort.

The New York station of the Central Railroad of New Jersey is at the foot of Liberty street. The ferry to Jersey City, known as the "Communipaw Ferry," was the first legally established ferry between Manhattan Island and the Jersey shore. It was erected at the foot of Communipaw avenue in 1661, William Jansen being licensed to take charge of it. Jansen at once endeavored to establish a monopoly, claiming that under his license every one was obliged to cross in his boat. On the other hand, the people claimed that he had violated his license by refusing to ferry certain parties. Jansen's answer was that he had never refused to ferry those who would pay. The Governor and Council neatly solved the problem by deciding against both parties—ordering the Sheriff to assist Jansen in getting his pay, and threatening Jansen with dismissal if he refused to ferry anyone who was willing to pay. The ferry was to be in operation Mondays, Wednesdays and Fridays, but for an extra compensation, "four guilders in wampum," Jansen was obliged to ferry any person at any time. For more than a century thereafter there is no record concerning the ferry, but it probably continued to be patronized. At the close of the Revolution, in 1783, Aaron Longstreet & Co. advertised that the ferry would, at 3 P. M., convey passengers to Communipaw for the stage for Newark, whence they could proceed by "the Excellent New York and Philadelphia Running Machines" to Philadelphia in one day. When, however, the enemy evacuated Paulus Hook, the line of travel swerved that way. and the Communipaw Ferry, falling into disuse, was not revived until the Central Railroad of New Jersey was extended from Elizabethport to Jersey City, the old flat-boat ferry being suddenly rehabilitated in all the improvements of a century of progress. The act authorizing the extension was passed in 1860, and the railroad was opened for travel August 1, 1864.

The ferry being the most southerly of those crossing to the New Jersey shore, a finer view of New York Harbor is had than from the boats of any other of the Jersey railroads.

At the Jersey City terminus is a depot of the best modern construction. Its high clock-tower and gables are conspicuous features of the river front. Arched corridors lead from the ferry-house to the waiting-rooms. The general effect of this room and its accessories is spacious, yet graceful. It is 60 feet from the tiled floor to the apex of the roof, held by ornamental rafters of iron. A sky-light and the windows provide perfect light and ventilation. The walls are of buff glazed imported brick; the tiling of the floor is marble. The building is lighted by electricity. A stair-case and gallery lead to the offices on the second floor.

On the further right hand corner of the waiting-room (entering from the ferry), is a luxuriously furnished ladies' parlor, while on the same side, but near the entrance from the ferry, is a smoking-room. Among the accessories is an excellent restaurant.

The train-shed boasts one of the finest sky-lights in the country. Some idea of the size of the shed may be gained from the statement that there are twelve tracks which will hold fifteen cars each, or 180 cars might start at once and carry away 12,600 people, seated, and another 2,000 might occupy the aisles and platforms; 190 trains arrive at and depart from the station every twenty-four hours, and if each one had the full number of cars that might stand under the roof of the car-shed there would be 2,850 cars come and go every day, capable of seating 199,500 passengers every twenty-four hours.

From this station straightaway an important branch runs to Newark. It traverses Bergen Neck through a cut, with stations at some of the most pleasant residential districts of Jersey City, crosses the Hackensack and Passaic and then enters Newark with stations at East Ferry and Ferry streets (in the heart of the manufacturing district) and at Broad street, the last being the most central of any railroad station in Newark. From Newark a branch runs to Elizabethport, Elizabeth and Roselle, connecting at the first-named for the Jersey coast and Pine resorts and Freehold, and at the second for places on the main line and its connections. The trains for the track of the New Jersey Jockey Club run part of the distance over the Newark branch and the balance over the Newark and Elizabeth branch.

The main line follows the shore of Bergen Neck and crosses Newark Bay to Elizabethport, where connection is made for Perth Amboy and intermediate stops, for all points on the New York and Long Branch Railroad (the famous resorts of the Jersey coast and the race-course at Monmouth Park); on the Freehold and New York Railroad, and on the Jersey Southern Railroad (Lakewood and Atlantic City).

From Elizabeth the main stem proceeds through a series of beautiful and thriving suburban villages and towns (Roselle, Cranford, Westfield, Fanwood, Netherwood, Dunellen) and the city of Plainfield to Bound Brook. To this point the road is four-tracked to accommodate the Philadelphia, Baltimore and Washington express trains run by the Central Railroad of New Jersey in connection with the Philadelphia and Reading and the Baltimore and Ohio Railroads, and also the trains of the Lehigh Valley Railroad, which come in near Roselle. Bound Brook is the junction-point for the Philadelphia, Baltimore and Washington trains. Two stations beyond is Somerville, the terminus of the suburban system and the junction for the South Branch Railroad, which runs to Flemington. From Somerville the main stem continues through the flourishing manufacturing settlement of Raritan and a number of small villages to Phillipsburg, sending at High Bridge a branch to Schooley's Mountain, Budd's Lake, Lake Hopatcong and a line of rich mines in the Jersey Highlands. Then, having crossed the Delaware, the railroad reaches the rich coal and iron fields of Pennsylvania via Easton, Bethlehem and Allentown to Scranton.

CHURCHES —*Methodist-Episcopal.*—There are one or more Methodist-Episcopal Churches or congregations in every place along the route of the main stem of the Central Railroad of New Jersey and its branches.

Protestant-Episcopal.—Bayonne, Bergen Point, Bound Brook, Cranford, Dover, Dunellen, Elizabeth, Flemington, Greenville, Newark, Phillipsburg, Plainfield, Roselle, Somerville.

Presbyterian.—Bayonne, Bloomsbury, Bound Brook, Chester, Cranford, Dover, Dunellen, Elizabeth, Flemington, German Valley, Lafayette, Newark, Phillipsburg, Plainfield, Rockaway, Roselle, Schooley's Mountain, Westfield.

Congregational.—Bound Brook, Chester, Elizabeth, Newark, Plainfield, Westfield.

Baptist.—Bayonne, Cranford, Drakesville, Elizabeth, Flemington, Lafayette, Newark, Plainfield. Westfield.

Reformed. — Annandale, Bayonne, Bergen Point, Bound Brook, Greenville, High Bridge, Lafayette, Lebanon, Minnisink, Newark, North Branch, Plainfield, Raritan, Rockaway, Somerville.

Lutheran.—German Valley, Glen Gardner, Newark, Phillipsburg.

United Presbyterian.—Bloomington, Newark.

Universalist.—Newark.

Seventh-Day Baptist.—Dunellen.

Roman Catholic.—Annandale, Bergen Point, Bloomsbury, Bound Brook, Chester, Cranford, Dover, Dunellen, Elizabeth, Flemington, High Bridge, Hurd Junction, Lake Hopatcong, Lebanon, Newark, Neshanic, North Branch, Plainfield, Phillipsburg, Raritan, Rockaway, Scofield Mine, Somerville, Westfield, White House.

AMUSEMENTS AND SPORT.—In nearly all the places in the suburban system of the Central Railroad of New Jersey are tennis, base-ball, bowling, bicycling and athletic clubs. Among these the New Jersey Athletic Club, with grounds at Bergen Point, is known far beyond the limits of its location and even of the State; and the Cresents, of Plainfield, stand high among base-ball clubs. The Argonautas, of Bergen Point, and the Tritons, of Newark, enjoy more than local reputation among rowing associations. The lay of the land throughout the whole suburban system is highly favorable to bicycling and to driving and riding. The roads are level and hard and kept in admirable repair. Special rates

NEW JERSEY ATHLETIC CLUB.

of carriage hire will be found under various places described below. The general rate is from $1.00 per hour for single, and from $2.00 per hour for double teams.

The principal summer sport is fishing. The black bass of Lake Hopatcong are famous for size and gameness. The same species can be caught at Branch Mills, easily reached from Elizabeth, Roselle, Cranford, Westfield and Fanwood; in the pond at New Market, not far from Netherwood, Plainfield and Dunellen; in the Raritan, near Bound Brook, Somerville and Raritan; and at Budd's Lake, Morris Pond and Green Pond. Fine trout are caught in the North Branch of the Raritan, especially at Naughright, and a good trout stream comes down the mountain at Flanders. With stocking and careful enforcement of the game laws, all the brooks on Schooley's Mountain could again become what they once were—noted trout streams. Pickerel of course abound in all the lakes, those of Green Pond being especially noted for their size.

Quails, partridges and woodcock abound through the mountains. Woodport and Budd's Lake are sporting headquarters in season. In the early fall many freshwater ducks congregate at Budd's Lake. There is excellent shooting for plover and doves at Bound Brook. Guides with dogs can be hired for from $1.50 a day upwards.

INDUSTRIES.—*Jersey City.*—Terminal point. Location of docks for shipments of heavy freight in car lots—Pig Iron, Anthracite Coal, Bituminous Coal, etc., etc. *Port Liberty.*—An adjunct to Jersey City. Shipping port for Anthracite and Bituminous Coal. *Jersey City, Henderson Street.*—Local delivery and shipping point for Jersey City proper. Sugar Refinery.

Newark Branch: *Lafayette.*—Zinc Works. Radiator Works. Stoves and Furnaces. Green-house Plant. *West Bergen.*—Steel Works. Chemical Works. Fire-works Manufactory. Horticultural Builders. *Newark.*—Brass Goods. Fourteen Breweries. Brewers' Supplies. Brick. Carriages and Carriage Ornaments. Acids. Corsets. Curled Hair. Pins. Thirteen Leather Manufactories. Four Paper Box Manufactories. Cement. Celluloid Goods. Harness Trimmings. Bottles. Oil. Filterers. Steel and Iron. Varnish. Zinc Oxide. Wagon Springs.

Corliss Engines and Machinery. Three large Fertilizer and Bone Black Manufactories. Barrels. Furniture. Bic cles. Binder Boards. Blue Stone Works. Boilers. Boots and Shoes. Manufacturers' Supplies. Wooden Boxes. Printing Presses. Sash and Blinds. Thread. Tinware. Tobacco. Tools. Water Motors. Hardware. Cigars. Maccaroni.

Main Stem: *Claremont.*—Oil Refinery. National Storage Co. *Pamrapo.*—Stove Foundry. *Centreville.*—Paint Works. Silk Mills. *Constable Hook* (reached by spur from Centreville).—Immense Oil Refineries. Chemical Works. Copper Works. *Port Johnson* (reached by spur from Bergen Point).—Shipping port for Anthracite Coal. *Elizabethport.*—Copper Works. Chemical Works. Fertilizer Works. Oil Works. Machinery and Castings. Stoves. Cordage. Rope and Binders' Twine. Building Paper. Building Felt. Furnaces. Machinery. Shafting and Pulleys. Singer Sewing Machine Co. Oil Cloth. Street Car Motors. Shear Works. *Elizabeth.*—Oil. Grease. Crockery. Rubber Goods. Sash, Blinds and Doors. Hats. Canned Goods. Two Breweries. *Cranford.*—Folding Mats. *Fanwood.*—Furs. Binder Boards. *Plainfield.*—Location of the Pond Machine Tool Co., Manufacturers of Railroad Tools. Two Printing Press Manufactories. Carpets. Oil Cloth and Clothing. Flour. Prepared Flour. *Dunellen.*—Clothing. *Bound Brook.*—Woolen Mills. Lubricating Oil. Hose. Paint. Compressed Air Pumps. Car Heating Apparatus. *Finderne.*—Horses. *Somerville.*—Building Brick. Clothing. Packed Pork. *Raritan.*—Raritan Woolen Mills. Cloth Mills. Foundry Forges and Machinery. Agricultural Implements. *North Branch.*—Milk. Live Stock. Agricultural Products. *White House.*—Peach Baskets. Milk. Live Stock. Agricultural Products. *Lebanon.*—Milk. Live Stock. Agricultural Products. *Annandale.*—Milk. Live Stock. Agricultural Products. *High Bridge.*—Milk. Live Stock. Agricultural Products. Iron Ore. Car Wheels, and Axles and Forgings. *Glen Gardner.*—Parchment Paper. *New Hampton Junction.*—Agricultural Products. *New Hampton.*—Agricultural Products. *Asbury.*—Agricultural Products. Flour and Feed. *Valley.* — Agricultural Products. Iron Ore,

Springtown.—Agricultural Products. Flour and Feed. *Phillipsburg.*—Cast Iron Pipe. Clay and Pulp. Pig Iron, Stoves, Sheet Iron, Bar Iron, etc. Stand Pipe.

The agricultural district extends from Raritan to Springtown, inclusive. Daily shipments of milk, particularly, are very large, and during the peach season shipments of this fruit are also very heavy.

South Branch (*Somerville to Flemington*).—Milk and Agricultural Products.

High Bridge Branch: *Califon.*—Granite. Lime. Peaches. *Vernoy.*—Lime. *Middle Valley.*—Creameries. *German Valley.*—Granite. Lime. *Chester.*—Iron Works. Peaches. *Hacklebarney.*—Iron Works. *Bartley.*—Turbine Wheels. Creameries. *Flanders.*—Fire Sand. *Cary's.*—Fire Sand. *Kenvil.*—Dynamite. *Lake Hopatcong.*—Ice. *Ogden Mine Railroad.*—Iron Ore. *Port Oram.*—Iron Ore. Silk Mill. *Dover.*—Mining Machinery. Silk Mills. *Rockaway.*—Iron Foundry. Rock Crushers. Knitting Mills.

HOTELS.—Key: Italics mean that, in the author's opinion, the hotel is first-class; ordinary type means that it is second-class; a dagger (†) means that the hotel is very good of its class. The rates quoted are per day.

ANNANDALE: Annandale Hotel, $1.50. ASBURY: American House, $1.50. BAYONNE: Bayonne Hotel, $1.50; Riverside Hotel, $1.50. BERGEN POINT: La Tourette House, $2.00–$3.00. BLOOMSBURY: Bloomsbury Hotel, $1.50. BOUND BROOK: Fisher's Hotel,† $1.50; Hotel Gaddis, $2.00. BUDD'S LAKE: Forest House, $2.50–$3.00. CALIFON: Union House, $1.50. CHESTER: Chester Hotel, $1.50. CLINTON (near Annandale): Union Hotel, $2.00. CRANFORD: Cranford Hotel, $1.50. DENMARK LAKE: Merritt Park House,† $2.00–$2.50. DOVER: Jolley's Hotel,† $2.00 (a noted old-fashioned hostelry); Park House, $2.00; Mansion House, $1.50. DRAKESVILLE: Drakesville Hotel, $1.50. DUNELLEN: Taylor's Hotel, $1.50. ELIZABETH: Sheridan House, $2.00. FLANDERS: Flanders Hotel, $2.00. FLEMINGTON: County Hotel, $2.00; Union Hotel, $2.00. GERMAN VALLEY: German Valley Hotel, $1.75. GLEN GARDNER: Glen Gardner Hotel, $2.00. HIGH BRIDGE: American, $1.00. LAKE HOPATCONG: *Hotel Breslin,*† $4.00–$5.00; Lake View House, $2.00 per day, $12.00–$14.00 per week;

American House, $2.00 per day, $12.00–$14.00 per week; Nolan's Point Villa,† $2.00 per day, $12.00–$14.00 per week (within 1 minute of Lake Hopatcong Station). MINNISINK: (See Lake Hopatcong). NETHERWOOD: *Hotel Netherwood.* NEWARK: Continental, $2.00–$3.00; Park Hotel, $2.00–3.00. NORTH BRANCH: North Branch Hotel, $1.50. PAMRAPO: Bayswater, $1.50–$2.00. PHILLIPSBURG: Central Hotel, $1.50–$2.00; Columbia House, $1.50. PLAINFIELD: City Hotel, $2.00; Force's Hotel, $2.00; Laing's Hotel, $2.00. PORT ORAM: Port Oram Hotel, $1.00. RARITAN: Raritan Hotel, $2.00. ROCKAWAY: Rockaway Hotel, $2.00; Central House, $1.00. ROSELLE: Van Court (building—to be ready March, 1890). SCHOOLEY'S MOUNTAIN: Heath House,† $2.50; Dorincourt,† $3.00–$4.00. SOMERVILLE: County House, $2.00; Ten Eyck House, $2.00; Commercial, $2.00. VALLEY: West Portal, $1.00. WESTFIELD: Westfield House, $2.00.

Single Fare

CENTRAL RAILROAD COMPANY OF NEW JERSEY.

NEW YORK to	Commutation Rates 1st Mo.	2d Mo.	3d Mo.	4th Mo.	5th Mo.	6th Mo.	7th Mo.	8th Mo.	9th Mo.	10th Mo.	11th Mo.	12th Mo.	Long Term Payments 3 Mos.	6 Mos.	12 Mos.	Trip Tickets Single Trip.	Excursion.	School Rates 1 Mo.	2 Mos.	3 Mos.	4 Mos.	5 Mos.	6 Mos.	7 Mos.	8 Mos.	9 Mos.	10 Mos.	11 Mos.	12 Mos.
Communipaw	[illegible]	[illegible]	[illegible]	[illegible]	[illegible]	[illegible]	[illegible]	[illegible]	[illegible]	[illegible]	[illegible]	[illegible]	[illegible]	[illegible]	[illegible]	[illegible]	[illegible]	[illegible]	[illegible]	[illegible]	[illegible]	[illegible]	[illegible]	[illegible]	[illegible]	[illegible]	[illegible]	[illegible]	[illegible]
Lafayette (Newark Br.)	[illegible]	[illegible]	[illegible]	[illegible]	[illegible]	[illegible]	[illegible]	[illegible]	[illegible]	[illegible]	[illegible]	[illegible]	[illegible]	[illegible]	[illegible]	[illegible]	[illegible]	[illegible]	[illegible]	[illegible]	[illegible]	[illegible]	[illegible]	[illegible]	[illegible]	[illegible]	[illegible]	[illegible]	[illegible]
Arlington Ave. (Newark Br.)	[illegible]	[illegible]	[illegible]	[illegible]	[illegible]	[illegible]	[illegible]	[illegible]	[illegible]	[illegible]	[illegible]	[illegible]	[illegible]	[illegible]	[illegible]	[illegible]	[illegible]	[illegible]	[illegible]	[illegible]	[illegible]	[illegible]	[illegible]	[illegible]	[illegible]	[illegible]	[illegible]	[illegible]	[illegible]
Jackson Ave. (Newark Br.)	[illegible]	[illegible]	[illegible]	[illegible]	[illegible]	[illegible]	[illegible]	[illegible]	[illegible]	[illegible]	[illegible]	[illegible]	[illegible]	[illegible]	[illegible]	[illegible]	[illegible]	[illegible]	[illegible]	[illegible]	[illegible]	[illegible]	[illegible]	[illegible]	[illegible]	[illegible]	[illegible]	[illegible]	[illegible]
West Bergen (Newark Br.)	[illegible]	[illegible]	[illegible]	[illegible]	[illegible]	[illegible]	[illegible]	[illegible]	[illegible]	[illegible]	[illegible]	[illegible]	[illegible]	[illegible]	[illegible]	[illegible]	[illegible]	[illegible]	[illegible]	[illegible]	[illegible]	[illegible]	[illegible]	[illegible]	[illegible]	[illegible]	[illegible]	[illegible]	[illegible]
Newark (Newark Br.)													[illegible]	[illegible]	[illegible]	[illegible]	[illegible]	[illegible]	[illegible]	[illegible]	[illegible]	[illegible]	[illegible]	[illegible]	[illegible]	[illegible]	[illegible]	[illegible]	[illegible]
Claremont	[illegible]	[illegible]	[illegible]	[illegible]	[illegible]	[illegible]	[illegible]	[illegible]	[illegible]	[illegible]	[illegible]	[illegible]	[illegible]	[illegible]	[illegible]	[illegible]	[illegible]	[illegible]	[illegible]	[illegible]	[illegible]	[illegible]	[illegible]	[illegible]	[illegible]	[illegible]	[illegible]	[illegible]	[illegible]
Greenville	[illegible]	[illegible]	[illegible]	[illegible]	[illegible]	[illegible]	[illegible]	[illegible]	[illegible]	[illegible]	[illegible]	[illegible]	[illegible]	[illegible]	[illegible]	[illegible]	[illegible]	[illegible]	[illegible]	[illegible]	[illegible]	[illegible]	[illegible]	[illegible]	[illegible]	[illegible]	[illegible]	[illegible]	[illegible]
Pamrapo	[illegible]	[illegible]	[illegible]	[illegible]	[illegible]	[illegible]	[illegible]	[illegible]	[illegible]	[illegible]	[illegible]	[illegible]	[illegible]	[illegible]	[illegible]	[illegible]	[illegible]	[illegible]	[illegible]	[illegible]	[illegible]	[illegible]	[illegible]	[illegible]	[illegible]	[illegible]	[illegible]	[illegible]	[illegible]
Bayonne	[illegible]	[illegible]	[illegible]	[illegible]	[illegible]	[illegible]	[illegible]	[illegible]	[illegible]	[illegible]	[illegible]	[illegible]	[illegible]	[illegible]	[illegible]	[illegible]	[illegible]	[illegible]	[illegible]	[illegible]	[illegible]	[illegible]	[illegible]	[illegible]	[illegible]	[illegible]	[illegible]	[illegible]	[illegible]
Centreville	[illegible]	[illegible]	[illegible]	[illegible]	[illegible]	[illegible]	[illegible]	[illegible]	[illegible]	[illegible]	[illegible]	[illegible]	[illegible]	[illegible]	[illegible]	[illegible]	[illegible]	[illegible]	[illegible]	[illegible]	[illegible]	[illegible]	[illegible]	[illegible]	[illegible]	[illegible]	[illegible]	[illegible]	[illegible]
Bergen Point	[illegible]	[illegible]	[illegible]	[illegible]	[illegible]	[illegible]	[illegible]	[illegible]	[illegible]	[illegible]	[illegible]	[illegible]	[illegible]	[illegible]	[illegible]	[illegible]	[illegible]	[illegible]	[illegible]	[illegible]	[illegible]	[illegible]	[illegible]	[illegible]	[illegible]	[illegible]	[illegible]	[illegible]	[illegible]
Elizabethport	[illegible]	[illegible]	[illegible]	[illegible]	[illegible]	[illegible]	[illegible]	[illegible]	[illegible]	[illegible]	[illegible]	[illegible]	[illegible]	[illegible]	[illegible]	[illegible]	[illegible]	[illegible]	[illegible]	[illegible]	[illegible]	[illegible]	[illegible]	[illegible]	[illegible]	[illegible]	[illegible]	[illegible]	[illegible]
Elizabeth	[illegible]	[illegible]	[illegible]	[illegible]	[illegible]	[illegible]	[illegible]	[illegible]	[illegible]	[illegible]	[illegible]	[illegible]	[illegible]	[illegible]	[illegible]	[illegible]	[illegible]	[illegible]	[illegible]	[illegible]	[illegible]	[illegible]	[illegible]	[illegible]	[illegible]	[illegible]	[illegible]	[illegible]	[illegible]
El Mora	[illegible]	[illegible]	[illegible]	[illegible]	[illegible]	[illegible]	[illegible]	[illegible]	[illegible]	[illegible]	[illegible]	[illegible]	[illegible]	[illegible]	[illegible]	[illegible]	[illegible]	[illegible]	[illegible]	[illegible]	[illegible]	[illegible]	[illegible]	[illegible]	[illegible]	[illegible]	[illegible]	[illegible]	[illegible]
Roselle	[illegible]	[illegible]	[illegible]	[illegible]	[illegible]	[illegible]	[illegible]	[illegible]	[illegible]	[illegible]	[illegible]	[illegible]	[illegible]	[illegible]	[illegible]	[illegible]	[illegible]	[illegible]	[illegible]	[illegible]	[illegible]	[illegible]	[illegible]	[illegible]	[illegible]	[illegible]	[illegible]	[illegible]	[illegible]
Cranford	[illegible]	[illegible]	[illegible]	[illegible]	[illegible]	[illegible]	[illegible]	[illegible]	[illegible]	[illegible]	[illegible]	[illegible]	[illegible]	[illegible]	[illegible]	[illegible]	[illegible]	[illegible]	[illegible]	[illegible]	[illegible]	[illegible]	[illegible]	[illegible]	[illegible]	[illegible]	[illegible]	[illegible]	[illegible]
Westfield	[illegible]	[illegible]	[illegible]	[illegible]	[illegible]	[illegible]	[illegible]	[illegible]	[illegible]	[illegible]	[illegible]	[illegible]	[illegible]	[illegible]	[illegible]	[illegible]	[illegible]	[illegible]	[illegible]	[illegible]	[illegible]	[illegible]	[illegible]	[illegible]	[illegible]	[illegible]	[illegible]	[illegible]	[illegible]
Fanwood	[illegible]	[illegible]	[illegible]	[illegible]	[illegible]	[illegible]	[illegible]	[illegible]	[illegible]	[illegible]	[illegible]	[illegible]	[illegible]	[illegible]	[illegible]	[illegible]	[illegible]	[illegible]	[illegible]	[illegible]	[illegible]	[illegible]	[illegible]	[illegible]	[illegible]	[illegible]	[illegible]	[illegible]	[illegible]
Netherwood	[illegible]	[illegible]	[illegible]	[illegible]	[illegible]	[illegible]	[illegible]	[illegible]	[illegible]	[illegible]	[illegible]	[illegible]	[illegible]	[illegible]	[illegible]	[illegible]	[illegible]	[illegible]	[illegible]	[illegible]	[illegible]	[illegible]	[illegible]	[illegible]	[illegible]	[illegible]	[illegible]	[illegible]	[illegible]
Plainfield	[illegible]	[illegible]	[illegible]	[illegible]	[illegible]	[illegible]	[illegible]	[illegible]	[illegible]	[illegible]	[illegible]	[illegible]	[illegible]	[illegible]	[illegible]	[illegible]	[illegible]	[illegible]	[illegible]	[illegible]	[illegible]	[illegible]	[illegible]	[illegible]	[illegible]	[illegible]	[illegible]	[illegible]	[illegible]
Grant Ave.	[illegible]	[illegible]	[illegible]	[illegible]	[illegible]	[illegible]	[illegible]	[illegible]	[illegible]	[illegible]	[illegible]	[illegible]	[illegible]	[illegible]	[illegible]	[illegible]	[illegible]	[illegible]	[illegible]	[illegible]	[illegible]	[illegible]	[illegible]	[illegible]	[illegible]	[illegible]	[illegible]	[illegible]	[illegible]
Evona	[illegible]	[illegible]	[illegible]	[illegible]	[illegible]	[illegible]	[illegible]	[illegible]	[illegible]	[illegible]	[illegible]	[illegible]	[illegible]	[illegible]	[illegible]	[illegible]	[illegible]	[illegible]	[illegible]	[illegible]	[illegible]	[illegible]	[illegible]	[illegible]	[illegible]	[illegible]	[illegible]	[illegible]	[illegible]
Dunellen	[illegible]	[illegible]	[illegible]	[illegible]	[illegible]	[illegible]	[illegible]	[illegible]	[illegible]	[illegible]	[illegible]	[illegible]	[illegible]	[illegible]	[illegible]	[illegible]	[illegible]	[illegible]	[illegible]	[illegible]	[illegible]	[illegible]	[illegible]	[illegible]	[illegible]	[illegible]	[illegible]	[illegible]	[illegible]
Bound Brook	[illegible]	[illegible]	[illegible]	[illegible]	[illegible]	[illegible]	[illegible]	[illegible]	[illegible]	[illegible]	[illegible]	[illegible]	[illegible]	[illegible]	[illegible]	[illegible]	[illegible]	[illegible]	[illegible]	[illegible]	[illegible]	[illegible]	[illegible]	[illegible]	[illegible]	[illegible]	[illegible]	[illegible]	[illegible]
Finderne	[illegible]	[illegible]	[illegible]	[illegible]	[illegible]	[illegible]	[illegible]	[illegible]	[illegible]	[illegible]	[illegible]	[illegible]	[illegible]	[illegible]	[illegible]	[illegible]	[illegible]	[illegible]	[illegible]	[illegible]	[illegible]	[illegible]	[illegible]	[illegible]	[illegible]	[illegible]	[illegible]	[illegible]	[illegible]
Somerville	[illegible]	[illegible]	[illegible]	[illegible]	[illegible]	[illegible]	[illegible]	[illegible]	[illegible]	[illegible]	[illegible]	[illegible]	[illegible]	[illegible]	[illegible]	[illegible]	[illegible]	[illegible]	[illegible]	[illegible]	[illegible]	[illegible]	[illegible]	[illegible]	[illegible]	[illegible]	[illegible]	[illegible]	[illegible]
[illegible] (So. Br. R.R.)	[illegible]	[illegible]	[illegible]	[illegible]	[illegible]	[illegible]	[illegible]	[illegible]	[illegible]	[illegible]	[illegible]	[illegible]	[illegible]	[illegible]	[illegible]	[illegible]	[illegible]	[illegible]	[illegible]	[illegible]	[illegible]	[illegible]	[illegible]	[illegible]	[illegible]	[illegible]	[illegible]	[illegible]	[illegible]
Flagtown (So. Br. R.R.)	[illegible]	[illegible]	[illegible]	[illegible]	[illegible]	[illegible]	[illegible]	[illegible]	[illegible]	[illegible]	[illegible]	[illegible]	[illegible]	[illegible]	[illegible]	[illegible]	[illegible]	[illegible]	[illegible]	[illegible]	[illegible]	[illegible]	[illegible]	[illegible]	[illegible]	[illegible]	[illegible]	[illegible]	[illegible]
Neshanic (So. Br. R.R.)	[illegible]	[illegible]	[illegible]	[illegible]	[illegible]	[illegible]	[illegible]	[illegible]	[illegible]	[illegible]	[illegible]	[illegible]	[illegible]	[illegible]	[illegible]	[illegible]	[illegible]	[illegible]	[illegible]	[illegible]	[illegible]	[illegible]	[illegible]	[illegible]	[illegible]	[illegible]	[illegible]	[illegible]	[illegible]
Woodfern (So. Br. R.R.)	[illegible]	[illegible]	[illegible]	[illegible]	[illegible]	[illegible]	[illegible]	[illegible]	[illegible]	[illegible]	[illegible]	[illegible]	[illegible]	[illegible]	[illegible]	[illegible]	[illegible]	[illegible]	[illegible]	[illegible]	[illegible]	[illegible]	[illegible]	[illegible]	[illegible]	[illegible]	[illegible]	[illegible]	[illegible]
Three Bridges (So. Br. R.R.)	[illegible]	[illegible]	[illegible]	[illegible]	[illegible]	[illegible]	[illegible]	[illegible]	[illegible]	[illegible]	[illegible]	[illegible]	[illegible]	[illegible]	[illegible]	[illegible]	[illegible]	[illegible]	[illegible]	[illegible]	[illegible]	[illegible]	[illegible]	[illegible]	[illegible]	[illegible]	[illegible]	[illegible]	[illegible]
Flemington (So. Br. R.R.)	[illegible]	[illegible]	[illegible]	[illegible]	[illegible]	[illegible]	[illegible]	[illegible]	[illegible]	[illegible]	[illegible]	[illegible]	[illegible]	[illegible]	[illegible]	[illegible]	[illegible]	[illegible]	[illegible]	[illegible]	[illegible]	[illegible]	[illegible]	[illegible]	[illegible]	[illegible]	[illegible]	[illegible]	[illegible]
Raritan	[illegible]	[illegible]	[illegible]	[illegible]	[illegible]	[illegible]	[illegible]	[illegible]	[illegible]	[illegible]	[illegible]	[illegible]	[illegible]	[illegible]	[illegible]	[illegible]	[illegible]	[illegible]	[illegible]	[illegible]	[illegible]	[illegible]	[illegible]	[illegible]	[illegible]	[illegible]	[illegible]	[illegible]	[illegible]
North Branch	[illegible]	[illegible]	[illegible]	[illegible]	[illegible]	[illegible]	[illegible]	[illegible]	[illegible]	[illegible]	[illegible]	[illegible]	[illegible]	[illegible]	[illegible]	[illegible]	[illegible]	[illegible]	[illegible]	[illegible]	[illegible]	[illegible]	[illegible]	[illegible]	[illegible]	[illegible]	[illegible]	[illegible]	[illegible]
White House	[illegible]	[illegible]	[illegible]	[illegible]	[illegible]	[illegible]	[illegible]	[illegible]	[illegible]	[illegible]	[illegible]	[illegible]	[illegible]	[illegible]	[illegible]	[illegible]	[illegible]	[illegible]	[illegible]	[illegible]	[illegible]	[illegible]	[illegible]	[illegible]	[illegible]	[illegible]	[illegible]	[illegible]	[illegible]
Lebanon	[illegible]	[illegible]	[illegible]	[illegible]	[illegible]	[illegible]	[illegible]	[illegible]	[illegible]	[illegible]	[illegible]	[illegible]	[illegible]	[illegible]	[illegible]	[illegible]	[illegible]	[illegible]	[illegible]	[illegible]	[illegible]	[illegible]	[illegible]	[illegible]	[illegible]	[illegible]	[illegible]	[illegible]	[illegible]
Annandale	[illegible]	[illegible]	[illegible]	[illegible]	[illegible]	[illegible]	[illegible]	[illegible]	[illegible]	[illegible]	[illegible]	[illegible]	[illegible]	[illegible]	[illegible]	[illegible]	[illegible]	[illegible]	[illegible]	[illegible]	[illegible]	[illegible]	[illegible]	[illegible]	[illegible]	[illegible]	[illegible]	[illegible]	[illegible]
High Bridge	[illegible]	[illegible]	[illegible]	[illegible]	[illegible]	[illegible]	[illegible]	[illegible]	[illegible]	[illegible]	[illegible]	[illegible]	[illegible]	[illegible]	[illegible]	[illegible]	[illegible]	[illegible]	[illegible]	[illegible]	[illegible]	[illegible]	[illegible]	[illegible]	[illegible]	[illegible]	[illegible]	[illegible]	[illegible]
Glen Gardner	[illegible]	[illegible]	[illegible]	[illegible]	[illegible]	[illegible]	[illegible]	[illegible]	[illegible]	[illegible]	[illegible]	[illegible]	[illegible]	[illegible]	[illegible]	[illegible]	[illegible]	[illegible]	[illegible]	[illegible]	[illegible]	[illegible]	[illegible]	[illegible]	[illegible]	[illegible]	[illegible]	[illegible]	[illegible]
Junction	[illegible]	[illegible]	[illegible]	[illegible]	[illegible]	[illegible]	[illegible]	[illegible]	[illegible]	[illegible]	[illegible]	[illegible]	[illegible]	[illegible]	[illegible]	[illegible]	[illegible]	[illegible]	[illegible]	[illegible]	[illegible]	[illegible]	[illegible]	[illegible]	[illegible]	[illegible]	[illegible]	[illegible]	[illegible]
Asbury																[illegible]	[illegible]	[illegible]	[illegible]	[illegible]									
Valley																[illegible]	[illegible]	[illegible]	[illegible]	[illegible]									
Bloomsbury																[illegible]	[illegible]	[illegible]	[illegible]	[illegible]									
Springtown																[illegible]	[illegible]	[illegible]	[illegible]	[illegible]									
Phillipsburg																[illegible]	[illegible]	[illegible]	[illegible]	[illegible]									
Easton																[illegible]	[illegible]	[illegible]	[illegible]	[illegible]									

NEWARK AND ELIZABETH BRANCH.

	Single Fare.	Excursion.
Elizabethport	$0 15	$0 25
Spring St.	15	[illegible]
Elizabeth	15	[illegible]
El Mora	20	[illegible]
Roselle	25	[illegible]

HIGH BRIDGE BRANCH.

	Single Fare.	Excursion.
Califon	[illegible]	[illegible]
[illegible]	[illegible]	[illegible]
Middle Valley	[illegible]	[illegible]
German Valley (Schooley's Mountain)	[illegible]	[illegible]
Naughright	[illegible]	[illegible]
[illegible]	[illegible]	[illegible]
[illegible] Lake	[illegible]	[illegible]
[illegible]	[illegible]	[illegible]
[illegible]	[illegible]	[illegible]
[illegible]	[illegible]	[illegible]
[illegible] Junction	[illegible]	[illegible]
Port Morris	[illegible]	[illegible]
[illegible]	[illegible]	[illegible]
[illegible]	[illegible]	[illegible]
[illegible] Junction	[illegible]	[illegible]
[illegible]	[illegible]	[illegible]
Lake Hopatcong	[illegible]	[illegible]
[illegible]	[illegible]	[illegible]
[illegible]	[illegible]	[illegible]
[illegible]	[illegible]	[illegible]
[illegible]	[illegible]	[illegible]
Chester	[illegible]	[illegible]
Chester Furnace	[illegible]	[illegible]

REAL ESTATE VALUES, HOUSE RENT AND BOARD.

(Prices given are in all cases for desirable plots, and in "House Rent" column for well built, modern houses, with all improvements, and desirably located.)

	Building Lots.	Factory Sites per acre.	Farms per acre.	House Rent per annum.
Newark......	$500 to $3,000	$2,000 and up.	.....	$300 and up.
Jersey City ..	500 and up.			
Lafayette....	500 "			
Arlington Av.	500 "			
Jackson Ave.	500 "	2,000 "	......	300 "
West Bergen.	250 "			
Communip'w	500 "			
Claremont....	....			
Greenville ...	250 "			
Pamrapo				
Bayonne.. ..	250 "	2,000 "	.....	250 "
Centreville...				
Bergen Point.				
Elizabethpo't	500 "	2,000 "	......	200 "
Spring St.....				
Elizabeth	1,000 "	....	......	
El Mora.. ..	200 "	2,000 "	......	300 "
Roselle.......	600 "	2,000 "	.....	
Cranford	200 "	1,000 "	$500	250 "
Westfield ...	200 "	1,000 "	500	
Fanwood	300 to 1,500	500 "	300	300 "
Netherwood .	300 to 1,000	500 "	......	....
Plainfield.....	300 and up.	3,000 "	.... .	300 "
Grant Ave....	250 "	2,000 "	......	200 "
Evona	200 "	2,000 "	.....	
Dunellen ...	200 "	....	......	150 "
Bound Brook.	250 "	....	......	
Finderne				
Somerville...	250 "	....	.....	250 "
Raritan	200 "	....	.....	150 "
North Branch				
White House.				
Lebanon.....				
Annandale...				
High Bridge..				
Glen Gardner	150 "	200 "	200 and up.	150 "
Junction				
Asbury				
Valley				
Bloomsbury..				
Springtown ..				
Phillipsburg..	500 "	....	......	250 "
Lake Hopatcong	500 to 3,090	....	.....	....

Board is from $10.00 a week upwards from Jersey City to Somerville; beyond that point and along the High Bridge branch, from $8.00 a week upwards.

CHAPTER I.

NEWARK

The Newark branch of the Central Railroad of New Jersey leaves the main stem at Communipaw and cuts through the ridge of trap rock, which forms the high ground of Bergen Neck, with stations at Lafayette, Arlington avenue, Jackson avenue and West Bergen; crosses the Morris Canal, the Hackensack and the Passaic rivers, and then speeds over the meadows to Newark, stopping at East Ferry street, Ferry street and Broad street, opposite the City Hall, the most central railroad station in Newark. The Broad street cars pass the door, and from Market street at the next corner to the North, other lines may be taken (see p. 17). From Newark the Central Railroad of New Jersey runs a branch *via* Elizabethport to Elizabeth, connecting at Elizabethport for all points on the New York & Long Branch Railroad, the Jersey Southern Railroad, the Freehold & New York Railroad and the Atlantic Highlands Railroad, and at Elizabeth for points on the main line, for Philadelphia, intermediate points on the Philadelphia & Reading Railroad, and for Baltimore, Washington and the West, *via* the Baltimore & Ohio

Newark with its population of 170,000, its vast industrial interests, which make it the ninth manufacturing centre of the country, its broad avenues, fine residential streets and, above all, its modern, progressive spirit, is the most important suburb reached by the Central Railroad of New Jersey. Its history well illustrates Emerson's remark to the effect that you find as many heroes in the counting-house as on the battle-field. For the story of Newark is one of adventure in the founding and developing of useful arts.

INDUSTRIES.—Industrial history when followed closely year by year makes pretty dry reading; but, when surveyed by epochs—read with the aid of glasses which bring perhaps twenty-five years into one focus—it presents astonishing and even romantic statistics. Newark's industrial records extend back almost to the date of its founding. Its people seem to have taken as naturally to manufacturing as ducks to water. Leather was made in Newark already in 1676, for the town records show the appointment of a sealer of leather. In 1683 Newark cider was as famous as Jersey apple-jack is now, Deputy Gov. Thomas Rudyard writing to a friend in London, that "at a place called Newark is made great quantities of Cyder exceeding any we can have from New England or Rhod Island or Long Island." In 1721 free-stone quarrying had become quite an industry. From then until the Revolution, Newark's progress was slow, but steady. After the Revolution its manufactories began to attract attention elsewhere. This was especially the case with shoe-making, which had naturally followed closely in the wake of tanning. In fact a tanner, Moses N. Combs, was the first shoe manufacturer on an extensive scale for export. This Combs was a character. He tanned souls as well as sole-leather; for he was a clergyman, and, for a while, being dissatisfied with the strict orthodox form of worship at the First Presbyterian Church he established a separate congregation.

By 1806 shoe-making had become so important an industry in Newark that, in a map of the town published that year, the figure of a shoe-maker at work was engraved in the emblematic design over a few descriptive

MORRIS CANAL—NEWARK

sentences, one of which says the town "is noted for its Cider, the making of carriages of all sorts, coach lace, men's and women's shoes. In the manufacture of this last article, one-third of the inhabitants are constantly employed." Carriage-making was then and is still one of Newark's most important industries.

The first establishment in the United States for the exclusive manufacture of jewelry was founded in Newark in 1801. The first iron foundry was established prior to 1810—the exact year cannot be determined—in Washington street, on the site now occupied by the Second Presbyterian church. Hatting and chair-making also flourished. In fact, already in the early part of this century, Newark was a lively manufacturing centre. A statement based upon the census returns of 1810 shows the value of articles manufactured in Essex Co. to have been $1,169,871, in which amount the boots, shoes and slippers made an item of $400,000. In 1834 Newark was made a port of entry, and in 1835 the imports were $2,500,000 and the exports $8,000,000. The first Collector of the Port was Archer Gifford, who kept the Gifford Tavern, quite a noted hostelry, especially among sportsmen. A fox-hunting scene was painted on the sign-board, and here the fox hunters were wont to gather after they had given Reynard a chase over the meadows.

In 1830 a local committee made a careful canvass of Newark's manufacturing industries, and in their report mention is made of two breweries. The beer-brewing industry, which has now attained enormous proportions in Newark, was then just starting in a small way. Subject, of course, to temporary checks, the industries of Newark have steadily developed. According to the last census (1880) there were 1,291 manufacturing establishments in Newark with a capital of $31,055,565, producing articles valued at $66,985,766, and employing 41,510 hands at a cost in wages of $14,784,388. According to Holbrook's Directory for 1889, there are now in Newark about 1,650 manufacturing establishments. Among these are 26 breweries, 49 makers of carriages and appurtenances, 5 thread works, one of which employs over 3,000 operatives, 87 clothing manufacturers, 21 boot and shoe factories, 23 brass foundries, 22 button factories, 50 hat manufacturers, 22 iron founders, 63 leather manufacturers and dealers, 2 macaroni factories, 111 segar manufacturers, 6 fertilizer works, 6 smelters and 10 tanners. The figures quoted give briefly but eloquently the history of Newark's industrial progress.

Now, a resident of Newark can clothe himself from head to foot in Newark-made garments, sewed with Newark-made thread; sit on a chair made in Newark, at a table made in Newark, spread with a cloth made in Newark and set with Newark-made china, glass and cutlery; and eat, with a fork made in Newark, vegetables grown under the stimulating influence of Newark fertilizers; smoke, after dinner, a cigar rolled in Newark, over a bottle of Newark-brewed beer; and, in his last moments, contemplate calmly the approach of death, because he knows that he will be borne to his grave in a Newark-made hearse, followed by Newark-made carriages, and buried in a grave dug with a Newark-made spade.

As an illustration of how thoroughly the soul of Newark is absorbed in manufactories, the following conversation, overheard at the sea shore last summer, may be quoted:

Mr. Beau.—"Are you acquainted with Scott's works, Miss Belle?"

Miss Belle (of Newark).—"Why, no! I never heard of them; are they on the Passaic?"

Of course the growth of Newark's population has been

FIRST FLOOR
THIRD FLOOR

commensurate with the expansion of its industrial interests. Following is part of the striking record:

1826... 8,017	1848... 30,000	1880...126,000
1830...10,995	1860... 71,941	1885...152,988
1837...20,079	1870...105,542	1889...170,000

SOCIAL HISTORY.—The social history of Newark is quite as interesting as its industrial record, especially to any one who can get beneath the facts at the philosophy. Newark began as an hierarchy as complete as any which flourished in Biblical days. Its founders were New England Congregationalists, and they early resolved that "none be admitted freemen or free Burgesses within our Town upon Passaick River * * * but such planters as are members of some or other of the Congregational churches." The pastor of the Congregational Church (afterwards the First Presbyterian or "Old First" Church) was chosen and paid by the town, and the church building was the public meeting-house. The early history of this church and the early history of Newark are, therefore, identical. The town continued to be thus governed until about 1720, when, the first settlers having died or ceased to have a controlling influence in public affairs, the religious tests of citizenship were abandoned.

SETTLEMENT.—Newark was settled about the middle of May, 1666, by Puritans from Milford, New Haven and Branford, Conn. The initial step was taken in 1661, the restoration of Charles II having aroused a fear in some of the colonists that their liberty of conscience would be interfered with. Through a committee, of which Robert Treat was the head, negotiations were opened with Governor Stuyvesant, but, before the Connecticut colonists decided to emigrate, England had possessed herself of the New Netherlands, and it was under Carteret's administration that Newark was settled, its founders purchasing their title direct from the Indians. Tradition has it that the first of the little band of New Englanders to set foot on Newark soil was Elizabeth Swaine, a fair Puritan maiden of eighteen summers, whose lover, Josiah Ward, gallantly secured her this honor. Elizabeth must have been an engaging young person, for that she was wooed at least twice is a matter of historical record. Josiah Ward was her first successful suitor, and to him she was married. After his death she married David Ogden, and their son, Josiah Ogden, became the founder of the first Episcopal church in Newark (p. 9).

The cost of the territory purchased by the settlers, which included the greater part of what is now Essex Co., was equal to about $750; the consideration being "50 double hands of powder, 100 bars of lead, 20 axes, 20 coates, 10 guns, 20 pistols, 10 swords, 10 kettles, 4 blankets, 4 barrels of beer, 2 pair of breeathes, 50 knives, 20 hoes, 850 fathoms of wampum, 2 ankers of liquor (say 32 gallons), or something equivalent, and three troopers' coates."

Robert Treat, the leader of the settlers, and who may justly be regarded as the founder of Newark, was an interesting figure in our colonial history. He was both a good civilian and a good fighter; a man who could preside at legislative councils and pick off a hostile savage with his rifle with equal judgment. Treat was born in England. We hear of him in Milford as early as 1640. At the battle of Bloody Brook he "made no less than 17 fair shots at the enemy, and was thereby as often a mark for them." When Sir Edmund Andros attempted to wrest Connecticut's charter from her Governor, Treat presided in the Assembly chamber, and it is believed to have been at his suggestion that the lights were suddenly extinguished, so that Capt. Wadsworth was enabled to slip out and secrete the precious document in the Charter Oak. But, though brave in the presence of the enemy, he is said to have been extremely bashful with members of the gentler sex, and it is related that his first wife, Jane Taff, was obliged to lead him up to a proposal of marriage by observing, as he was dancing her on his knee (which was "permissable by their disparity of age and long intimacy"): "Robert, be still that; I had rather be Treatted than trotted!" Treat remained in Newark only some six years. The "Old First" church stands upon a portion of his "home lotte."

The name Newark is derived from New-Worke, for thus the settlement was called by Treat and others of his associates. The present name was substituted by Rev. Abraham Pierson, the first pastor of the new town, a native of Newark-on-Trent; this place also, it may be interesting to note, derived its name from "New-Work." Newark-on-Trent, 773 years old, has 14,000 inhabitants; Newark on the Passaic, only in the beginning of its third century, has a population of 170,000.

At the time of the original settlement, and for many years afterwards, wolves and bears were so numerous in the neighborhood of Newark that the town offered premiums for killing them, and one of the settlers added considerably to his possessions by establishing a wolf-pit.

THE "OLD FIRST."—The First Presbyterian Church of Newark, whose early history and that of Newark go hand in hand, may be said to have ante-dated even Newark itself, and to be the oldest English congregation in the State, though the first church structure in New Jersey was the Dutch Reformed of Bergen.

The "Old First" existed as a congregation already in Branford, Conn., in 1644, whence, in 1666, it was simply translated to Newark, where, in 1668, a place of worship was erected. The site was on the west side of Broad street, nearly opposite the present structure. In 1675, it served also as a public house of refuge, the inhabitants, alarmed at the Indian atrocities in New England, fearing an attack from the Hackensack natives. The "meeting house" was fortified and flanked with palisades. But the fears were groundless, the fairness of the settlers toward the Indians, in purchasing the land of them, proving on this occasion, as well as in the future, a guarantee of peace. There is no record of any disturbance between the people of Newark and the Indians.

Nothing, perhaps, illustrates better the spirit of fairness which governed the people of Newark in their dealings with others than the manner in which the boundary line between Newark and Elizabeth was settled. Worthies from both towns met on what has since been known as Divident Hill, near Bound Creek, and the proceedings opened and closed with prayer; the agreement being reached amicably and solemnly. They were just, but not weak; for they exacted with quiet dignity for themselves the justice which they meted out to others. When

Sir Edmund Andros sought to wrest New Jersey from Carteret, and issued a proclamation to that effect, Newark replied calmly but firmly: "The Town being met together, give their positive answer to the Governor of York's writ, that they have taken the oath of allegiance to the King, and fidelity to the present Government, and until we have sufficient order from his Majesty we will stand by the same."

The first pastor of the "Old First" Church was Rev. Abraham Pierson, of Nottinghamshire, England, and a graduate of Cambridge. He had ministered in Branford some 23 years before he removed to Newark with the Branford settlers, in the fall of 1666. He was a zealous Indian missionary, acquiring during his labors a sufficient knowledge of the language of those among whom he worked to compile a catechism in their tongue. This work was printed in 1660. He died in Newark, August 9, 1678, his son Abraham, who had been his assistant since 1672, succeeding him. The younger Pierson seems to have been popular, for he is described as "a fleshy, well-favoured and comely-looking man." In 1692, he severed his connection with the church, because he preferred a moderate Presbyterianism to the strict Congregationalism in which he was required to minister, and returned to Connecticut. When Yale College was founded, in 1701, he was chosen its first Rector or President, and was held in such esteem that the college was temporarily established at Killingworth, where he was pastor, to suit his convenience, and because of the love which his flock bore him.

He was succeeded in Newark by Rev. John Prudden. Meanwhile, somewhat of a change had come over the spirit in which the affairs of the town were administered. It was no longer a strictly ecclesiastical government, and the Congregational tests of citizenship were not so strictly enforced. Nearly all the original settlers were dead. Mr. Prudden himself seems to have been opposed to the mingling of politics and religion, and, during his seven years pastorate, the ecclesiastical spirit was probably still more eliminated from the administration of public affairs. With the appointment of his successor, Rev. Jabez Wakeman, the separation became more clearly defined, for the expense of his maintenance was not provided by a levy on the town, as in the case of his predecessors, but by voluntary subscription. Wakeman died in 1704, shortly after completing the fifth year of his pastorate, and for a period of five or six years following there was no regular pastor. Finally, an emissary was dispatched to Connecticut, the "great clerical hive," where he secured Rev. Nathaniel Bowers, who was pastor for six years, until his death in 1716. During his pastorate, probably about 1715, a new church building of stone, 40 feet square, with a steeple and bell, was erected a little to the north of the first meeting house.

A Mr. Buckingham succeeded Bowers for a short time, and then Rev. Joseph Webb was ordained by the Presbytery of Philadelphia, October 22, 1719. His Presbyterian ordination was an important incident in the history of Newark, for it marked a further drifting away from the original Congregational theory of town and church government. The character of the population had been somewhat changed by Scotch and Dutch accessions. In 1727, we learn of a Dutch church organization in Newark—the town whi h had in 1666 limited citizenship to Congregationalists—and it is thought that Episcopal services were held there as early as 1729. The first Episcopal Church of Newark—Trinity—was founded, as a result of a quarrel between Col. Josiah Ogden and the First Presbyterian Church. About the fall of 1733, he shocked the staid members of the Presbyterian congregation, of which he himself was a member, by harvesting his wheat on Sunday, in order to save it from the rain. The result was a bitter controversy, which ended in a schism, Col. Ogden and a number of his friends withdrawing from the "Old First" Church and founding Trinity.

The history of the "Old First" and of Newark now brings us to two characters, one of whom at least played a conspicuous part in the military history of the Revolution and in the civil history of the new-born republic. These are Aaron Burr and his son and namesake; the former a man of intellectual poise, dignity of bearing and weight of character; the latter inheriting his father's mental gifts, but uniting with them a passionate temperament which caused them to be misapplied, and made him, while one of the most brilliant and fascinating figures in American history, also one of the most contemned.

Rev. Aaron Burr was the seventh pastor of the First Presbyterian Church. He preached there when but 19 years old. During his pastorate he devoted much thought and energy to educational matters, and he was largely instrumental in founding the College of New Jersey (Princeton). In May, 1744, David Brainerd, the Indian missionary, was ordained in the "Old First." For a trifling indiscretion he had been expelled from Yale College and, although backed by Burr and other prominent divines, he was not restored. This and his subsequent ordination in Newark led to a synodical separation, which resulted in the establishment of the College of New Jersey as a rival to Yale. It was founded by converting Rev. Jonathan Dickinson's classical school at Elizabeth Town into a college. But when, soon after the opening of the college (May, 1747), Dickinson died, the students were placed under the care of Aaron Burr, at Newark, who, in November the year following, was elected President. The college remained in Newark eight years, when it was moved to Princeton (1756), where Burr died a year later. He had resigned his pastorate in 1755 in order to give his whole time to the affairs of the college. During his ministry a complete separation of his church from the town government was effected, a distinct act of incorporation being secured for the church.

During the early years of Burr's pastorate two riots occurred in Newark. They grew out of the claims of the English Proprietors that they and their predecessors alone could give legal titles, whereas the descendants of the settlers maintained that they had secured valid titles from the Indians themselves direct. The rioters twice broke open the gaol and liberated those who had been confined therein at the instance of the Proprietors, but there does not appear to have been any bloodshed.

Aaron Burr, the second, was born in Newark, February 6, 1756, in the parsonage, a fine stone mansion on the west side of Broad street, 34 feet south of what is now the southwest corner of William. In the yard stood four large trees which, when the house was torn down in 1835, were transplanted to Broad street, south of Pennington, where they are now flourishing. To this house, Aaron Burr, the elder, being very popular,

THE "OLD FIRST"—1668.

SECOND AND PRESENT BUILDING.
(" Old First.")

came many of the young people of Newark and its vicinity that they might have the nuptial knot tied by the famous divine, and it is probable that in no other house in New Jersey at that time were so many people made happy—or miserable. Aaron Burr, the second, was not sufficiently identified with Newark to render a detailed account of his career necessary. His elevation to the Vice-Presidency, his intrigue against Jefferson, the Blennerhassett incident, his duel with Hamilton, are matters well known to all who are acquainted with our national history.

After Aaron, the elder, removed to Princeton, Rev. John Brainerd, a brother of the Indian missionary, became pastor of the First Church. His successor in 1759 was Rev. Alexander Macwhorter, who had graduated at Princeton and completed his studies at Freehold under the famous William Tennent. Dr. Macwhorter was an earnest patriot and was obliged to flee when the British entered Newark. They looted the parsonage, destroying many valuable church records. He became a chaplain in Washington's army and participated in the council of war which decided upon the memorable crossing of the Delaware.

Before the Revolution already it had been decided to erect a new church edifice. Trenches for the foundations had been dug and metal for the bell collected. When the war broke out the metal was carted to what is now known as Maple Island and buried. In September, 1787, the foundations of the present edifice were commenced, and January 1, 1791, the church was dedicated. In the tower, just over the entrance, a tablet was inserted bearing this inscription attributed to William Peartree Smith, then treasurer of the society:

Aedem hanc amplissimam cultui divino dicatam, ex animo religioso et munificentia valde praeclara, Nov Arcae habitantes, cura sub pastorali Rev. Alexandri Macwhorter, S. T. D., primum qui posuit saxum, construxerunt, anno salutis, 1787; *Amer. Reipub. Foederatae* 12. *Auspicante Deo, Longem Perduret in Aeum.*

Dr. Macwhorter's pastorate terminated with his death, July 20, 1807. His successors have been Edward D. Griffin, till 1809; James Richards, till 1823; William T. Hamilton, till 1834; Ansel D. Eddy, till 1848; Jonathan

F. Stearns, till 1883; David R. Frazer, the present pastor. Dr. Macwhorter's study chair, an old-fashioned piece of furniture with a broad wooden shelf attached to the right arm is preserved in the rooms of the New Jersey Historical Society (p. 16).

The "Old First" has played so important a part in the history of Newark, and still wields so benign an influence, that it has been deemed advisable to give illustrations of the three buildings which the congregation has occupied since it settled on the Passaic, over two centuries ago. That of the first meeting house is from a drawing found in the corner of an old map, entitled "Our Towne on Passayke;" that of the second is enlarged from another old map. The third shows the exterior of the present structure, which stands on the east side of Broad street, just south of the Jersey Central station. The interior is not so old fashioned as one would expect from the history of the church. The decorations consist of elaborate mouldings. There is a large platform for the pulpit; a colonnade of steps leads up to it on either side; above it is an arch which half conceals a dome.

The church which founded Newark, and whose early history is identical with that of the city, is as flourishing as the city which it founded. It owns valuable real estate on Broad street, and enjoys also the support of a large and wealthy congregation. In the rear of the church is the grave-yard. North and west it is flanked by the rear of factories and on the south by the railroad. The whirring of machinery, the puffing of engines and the unkempt look of the grounds, arouse a feeling of resentment at the neglect which has fallen upon this spot, which once, doubtless, was tranquil and beautiful. The original burying ground was located in the rear of the engine-houses on the opposite side of Broad street. During the summer of 1889 the remains and headstones were removed to a vault in Fairmount Cemetery.

The First Presbyterian church is a venerable structure for this young country, and, with its traditions reaching back even beyond the settlement of Newark, it may be regarded as a fit monument to the sturdy band of pioneers whose spirit of pluck and perseverance seems to have been inherited by the community they

COCKLOFT HALL.

founded. Upon this spirit, as upon a foundation, the city of Newark has been built.

HISTORICAL INCIDENTS.—The Revolutionary history of Newark is devoid of picturesque details, the British having made Elizabeth the base of their operations in this section of New Jersey. In 1776, Washington, then on his retreat to the Delaware, passed through Newark, which has one claim to distinction over all other towns in which Washington and his troops were quartered during the Revolution—it has no Washington's Headquarters. Washington was in Newark five days, and, from his record as a sleeper elsewhere, we may be sure that he slept in some structure in Newark. But, unfortunately, it cannot be identified. It has been claimed that he stopped at the Gouverneur mansion—Irving's "Cockloft Hall"—situated back from Mt. Pleasant avenue and facing Gouverneur street; also at the old Eagle Tavern, which fronted on Broad street a little north of the present City Hall; and at the house of one of his officers, Capt. Huntington, which stood on the southeast corner of Broad street and Eighth avenue.

Washington entered Newark the evening of November 22, 1776, with his retreating forces. Six days afterwards, just as the Americans moved out, Cornwallis's vanguard moved in from the north. The English started in pursuit about December 1, leaving a strong guard in Newark, whose presence stimulated many who were Tories at heart to come out in their true colors. Newark, did not, however, suffer during the Revolution as did Monmouth, nor is her Revolutionary history so romantic as that of the coast counties. Her trials were confined to a few depredations directed against those who sympathized with the Americans.

An amusing incident in the history of Newark was the election held early in 1807 to determine upon the site for a new court-house building. At that time Elizabeth Town was part of Essex Co. and a rival claimant of Newark for the court-house site. In those days women had the right of suffrage, and they and the children for several days before the election did nothing but write ballots. On the day of election, the voters, men and women, were driven hurriedly from poll to poll, and voted as often as they could, the women vying with the men in ballot-box stuffing. So determined was the contest, that Gov. Pennington himself conducted "a strapping negress" to the polls and "joined her in the ballot." Men who had voted disguised themselves as women and voted over again, and boys also attired themselves as women and gained access to the polls. The result was in favor of Newark, but the frauds were so palpable—Newark's vote being nearly equal to her whole population—that the election was set aside, while the frauds committed by the women created such a scandal that the right of suffrage was taken away from them.

HISTORIC BUILDINGS.—The three most interesting historical buildings in Newark are the "Old First" and Trinity churches (pp. 7 and 20), and the old Gouverneur mansion, famous as Washington Irving's "Cockloft Hall," because of his frequent sojourns there and his references to it in "Salmagundi." At that time it was owned by Gouverneur Kemble, one of Irving's intimates. Pierre Irving, in his "Life and Letters of Washington Irving" says:

"Among Irving's associates at this time were Peter and Gouverneur Kemble, Henry Brevoort, Henry Ogden and James K. Paulding, who, with himself, his brother Peter and a few others, made up a small circle of intimates, designated by Peter as the 'Nine Worthies,' though Washington described them as 'The Lads of Kilkenny.' One of their resorts was an 'old family mansion' * * * which was on the banks of the Passaic, about a mile above Newark. * * * * It was full of antique furniture, and the walls were adorned with old family portraits. The place was in charge of an old man, his wife and a negro boy, who were its sole occupants, except when the nine, under the lead, and confident in the hospitality of the Patroon, as they styled its possessor, would sally forth from New York and enliven its solitude by their madcap pranks and juvenile orgies."

On the place was a summer-house and a fish-pond, of which Irving, to illustrate the peculiarities of the mythical Cockloft, says:

"An odd notion of the old gentleman was to blow up a large bed of rocks for the purpose of having a fish-pond, although the river ran at a distance of about a hundred yards from the house and was well stored with fish; but there was nothing, he said, like having things to one's self. And he would have a summer-house built on the margin of the pond; he would have it surrounded by elms and willows, and he would have a cellar dug under it for some incomprehensible purpose, which remains a secret to this day."

This summer-house as it was in 1859 was sketched by William A. Whitehead, who presented the drawing to the New Jersey Historical Society. Writing to this society not many years before his death, Irving says: "With Newark are associated in my mind many pleasant recollections of early days, and of social meetings at an old mansion on the banks of the Passaic."

The summer-house was demolished when Passaic street was extended. The mansion still stands on Mount Pleasant avenue, corner of Gouverneur street, but it is much altered, and has no relics of the days when Irving and his companions had their frolics there.

Modern Newark.—*City Hall*, northwest corner of Broad and William streets; *Police Headquarters*, 13 William street; *Post Office*, northwest corner of Broad and Academy streets; *Custom House*, same; *Court House*, intersection of Market and High streets.

Telegraph: *Newark District Telegraph Office*, 182 Market street; *Western Union Telegraph Co.* (*with Public Telephone*), 180 Market street; *Telegraph Stations*, in all the railroad stations.

Railroad Stations: *Central Railroad of New Jersey*, East Ferry street, Ferry street (northeast corner of Prospect street), Broad street (between Mechanic and Fair streets), opposite City Hall, a short distance from the Court House and Post Office — the most centrally located railroad station in Newark: *Pennsylvania Railroad*, Centre street, Market street (New Jersey Railroad avenue), Chestnut street (New Jersey Railroad avenue), Emmet street (New Jersey

Railroad avenue); *Delaware, Lackawanna & Western Railroad,* Broad street (Morris and Essex Railroad avenue); *Newark and Paterson Railrord* (*New York, Lake Erie & Western Railroad*), Ogden street, between Third and Fourth avenues, Grafton avenue and Oraton street; *New York & Greenwood Lake Railroad,* Verona and Washington avenues.

Newark Academy, 536–548 High street. This is the oldest of the many public schools of Newark, having been founded in 1775. It is a collegiate preparatory school.

Hospitals: *Newark City Hospital,* 116 Fairmount avenue (has also training school for nurses); *St. Michael's Hospital,* High street corner Central avenue; *Hospital of St. Barnabas,* 681 High street; *Newark German Hospital,* corner Bank and Wallace streets; *Hospital for Women and Children,* South Orange avenue near Bergen street; *Home for the Friendless,* same; *St. James R. C. Hospital,* corner Lafayette and Madison streets.

Miscellaneous: *Society for the Prevention of Cruelty to Children for Essex Co.,* 144 Market street; *Boys' Lodging House,* same; *Society for the Prevention of Cruelty to Animals,* 46 Fair street; *Board of Trade,* 764 Broad street; *Y. M. C. A.,* Clinton street near Broad; *Y. M. Cath. A.,* 76 and 78 New street; *Y. M. H. A.,* 30 Plane street; *Miner's Theatre,* 193 Market street.

Restaurant: G. Munzer & Co., one door north of Broad street station, Central Railroad Co. of New Jersey.

New Jersey Historical Society, northwest corner Broad and Bank streets, publishes "New Jersey Archives" (Colonial Records), and "Collection of New Jersey Historical Society" (papers relating to the history of the State), owns the original of the map "Newark or Pesayak Towne" (1666–1680), Dr. McWhorters study-chair (p. 11) and cane, coat and *chapeau* of Capt. James Lawrence—("Don't give up the ship!")—a stone from the house of Columbus in San Domingo, a piece of the Charter Oak, a piece of the dock of Delft Haven from which the Mayflower sailed in 1620, Robert Fulton's plans and drawings, and several Indian relics.

Newark Free Public Library, West Park street near Broad, opened 1889, with 10,000 volumes. The building, of granite, with halls and stairways in marble, is one of the finest structures in the State. The main library room will accommodate 200,000 volumes. There are, besides the catalogue room, with printed catalogues arranged according to subjects on tables and card catalogues in desks arm-high around the room, a reference library, a prettily-furnished reading-room for women, a main reading-room, and a class-room to be used by children of the public schools brought thither by their teachers. The library is supported by a tax levy at the rate of one-third of a mill on the dollar of taxable valuation. Books may be taken out on presentation of a card. Cards, good for three years, will be issued on a "directory identification." If the applicant is known, or his or her name is in the directory, a card will be given; if neither of these conditions exist, the card will be granted only after four days' delay. The four days will be used in ascertaining, by library messengers, the correctness of the name and address ot the applicant. Armed with the card, there is no further preliminary necessary—the library will be open to the applicant. Books may be kept two weeks, and a penalty will be exacted for keeping them longer—two cents a day. A postal-card notice will be sent at the expiration of a fortnight. If the book is not returned a week later, a library messenger will go to the house, obtain the book and collect twenty cents messenger service.

HORSE-CAR LINES.—Essex Passenger Railroad Co.—*Orange Line* (green car and green signal light): From Market Street Depot Pennsylvania Railroad Co., through Market, Broad and Orange streets; through Roseville and East Orange to Lincoln avenue, Orange. *Broad Street Line* (red car and red signal light): From Badger avenue, through Clinton avenue past Lincoln Park; through Broad street; through Belleville avenue to Mt. Pleasant Cemetery. *Belleville Line:* From Washington avenue to Belleville. *Roseville Line:* From Roseville Depot, through Warren street, Wallace place, Bank street past Court House; through Market, Bowery and Chapel streets and Albert avenue to Lockwood street. *Newark and Bloomfield Line* (yellow car and yellow signal light): From Station, foot of Broad street, through Broad and State streets, Summer and Bloomfield avenues to Bloomfield. (Branch through Mt. Prospect avenue to Old Bloomfield road.) *Harrison Line* (white car and blue signal light): From Davis avenue, Harrison; through

Harrison avenue, Bridge, Broad, Market, Union, Elm and Pacific streets to Pennington street.

Elizabeth and Newark Horse Railroad Co.—From Fourth avenue, through Ogden, Front, South Front and Mulberry streets, New Jersey Railroad avenue to and through Thomas street to Pennsylvania avenue; through Pennsylvania avenue to Miller street; through Miller street and Elizabeth avenue to City Line; thence to Waverly and Elizbaeth; connecting with East Newark, Irvington, South Orange, Orange, Bloomfield, Belleville and Roseville cars; and with Delaware, Lackawana & Western, Erie, Pennsylvania, and Central Railroad of New Jersey at various depots along line; also with all excursion boats on the Passaic River.

Newark and South Orange Horse Railroad Co.—*Newark and South Orange Line* (blue car and red signal light): From Market Street Depot, through Market street, past Court House; through Springfield and South Orange avenues, passing Fairmount and Holy Sepulchre Cemeteries, Shooting Park, Insane Asylum and Seton Hall College to South Orange.

Newark and Irvington Street Railway Co.—*Newark and Irvington Line* (yellow car and yellow signal light): From Market Street Depot, through Market street, past Court House; through Springfield avenue, past Woodland Cemetery, to Irvington.

CAB SERVICE.—*Standard Cab Co.*, 19 Division place; telephone No. 369; can be ordered by telephone from New York, Elizabeth, Paterson and any other places in telephonic connection with Newark. Rates:

Cabs—By the trip, one mile or fraction thereof, each passenger	$0 25
By the hour, first hour or fraction thereof,	1 00
Each succeeding hour	75
Coupés—By the hour only, first hour or fraction thereof	1 25
Each succeeding hour	1 00

CHURCHES.—For First Presbyterian Church, see p. 7. The Second Presbyterian Church of Newark was dedicated September 30, 1810; the Third in 1824. There are now nineteen churches of this denomination and two United Presbyterian churches in Newark.

The First Baptist (Peddie Memorial) Church was

organized June 6, 1806. Newark has now fourteen churches of this denomination.

The Peddie Memorial, in memory of Thomas B. Peddie, is the finest modern church structure in Newark. The general ground plan covers the entire lot, 100x107 feet, forming a circular floor plan.

Four prominent alcoves are arranged by a system of Roman arches which are furred and broken in the main domed lines of the arch that spans the great ceiling. In each of the four alcoves the galleries are slanted, and the seats are recessed back to the curved walls. Directly in the corner of Fulton and Broad streets and immediately above the porch is the memorial bay alcove, the window openings made to suit the memorial windows formerly in the chancel of the Academy street church.

The ceiling is formed in part by the four Roman arches on either side of the building, from the intersection of which massive groins run up into the great dome, the upper part of which is a huge stained-glass lantern of dome shape. The groins are heavy oak timbers, the panels of the ceiling of light wood, while metal ornaments in broad bands run horizontally around the domed ceiling, relieving the wood-work from monotony. A series of stained-glass dormer windows also encircle the ceiling of the lofty dome.

The center of the ornamentation and architectural effect is designed about the marble baptistery, which is placed in the center of the building; in other words, immediately back of the pulpit. Radiating from this point the seats gently curve, surrounding the pulpit and lectern, thus forming an amphitheatre. The baptistery is of solid carved marble, elevated several feet above the line of the main floor, and curved in general outline, with approaches corbelled and arched on the two sides, forming a cluster of grouped columns and arches, from the robing rooms, which are placed directly underneath the two organs on the north and south sides of the pulpit.

Directly behind the pulpit on the second story of the Fulton street wing of the building and facing the audience are the Sunday-school rooms.

Dividing the auditorium from the Sunday-school rooms is a rich and elaborate screen of cherry and wrought metal, with sliding partitions, so that the two general compartments of church and Sunday-school may be opened into one great audience room, when desired, capable of seating more than three thousand persons. When thus thrown open, the Sunday-school rooms form a richly alcoved chancel, and its groined ceiling and beautiful windows, as seen through the interstices of the metal screen, are an attractive architectural feature. One can readily imagine the effect when, during some solemn celebration, the screens suddenly part and disclose the children singing, with uplifted faces, their hymn of praise.

Of the eighteen Methodist Episcopal churches in Newark, Wesley Chapel—Halsey Street M. E. Church—is the oldest. The first meetings of the society were held in a bark mill which stood a few hundred yards from the present site of the Halsey street church, which was dedicated January 1, 1809.

There are twelve Episcopal churches in Newark. The oldest of these is *Trinity*. Here Washington, Robert Morris, Robert Livingston and Gen. Lord Sterling are known to have worshipped. The base of the steeple of the original structure, erected 1743–44, supports the steeple of the present edifice, which was consecrated May 21, 1810. The head-stone which marked the grave of Col. Ogden, Trinity's founder, in the old Presbyterian grave-yard, now forms part of the floor in the porch of Trinity, but the record is totally obliterated. The interior has an old-fashioned air, owing chiefly to the fact that the old pews have been retained. Those at the head of the main aisle with their bowed fronts are particularly quaint looking. There is a great high pulpit with narrow, winding stairs. There is a fine stained-glass window by Tiffany—The Baptism of Christ—in the rear of the chancel, and to the left the Gifford memorial window executed in England. The old burying ground is on Rector street, near by.

The first of the (Dutch) Reformed churches, of which there are nine, was organized in 1833.

The first Roman Catholic congregation met in 1826 in the basement of a private house. The first church record bears date November 3, 1829, and soon afterwards a small church was erected on the present site of St. John's Roman Catholic Church. St. Patrick's Cathedral was founded in 1850. There are now seventeen churches of this denomination in Newark, besides numerous parochial schools (6,699 pupils) and benevolent institutions, including several hospitals, and monasteries and fifteen convents, among which is St. Dominic's, whose inmates are engaged in the "Perpetual Adoration of the Most Blessed Sacrament," not being allowed beyond the confines of the convent except in cases of the direst necessity.

Besides these churches there are 2 Congregational, 1 Reformed Episcopal, 1 Methodist Protestant, 4 Lutheran, 1 Universalist, 2 Swedenborgian, 1 Reformed Catholic and 8 independent churches, and 4 Jewish Synagogues.

PARKS.—*Lincoln* (p. 24), junction of Clinton avenue and Broad street; *Military* (p. 25), Broad street and Park place, noted for its avenues of stately elms. Near the south end is a statue of Gen. Phil. Kearney. The park is on the site reserved by the settlers as a practice ground for the militia. *Washington* (p. 25), at the junction of Broad and Washington streets, on ground reserved by the settlers for a market place.

CEMETERIES.—*Mt. Pleasant,* Belleville avenue and Harney street, the principal cemetery of Newark, sloping down to the river and beautifully laid out; *St. John's,* Belleville avenue, near Fourth avenue; *Woodland,* Rose street, near Eighteenth avenue; *Fairmount* (p. 12), South Orange avenue.

CLUBS.—*Essex* (organized, 1876; incorporated, 1881), occupies a fine mansion on Park place, overlooking Military Park. Recently added to the building have been a spacious dining hall and billiard room. An air of old-fashioned comfort pervades the original portions of the house. Situated in one of the finest residential quarters of Newark, the club includes in its membership some of the oldest and wealthiest families in the State.

North End (organized, 1887), situated at the northwest corner of Broad street and Third avenue, the fine building (see illustration) having been erected for the Club. Special features of the interior arrangements are the roomy entrance hall, with oaken staircase, and broad, low fireplace; the reception room and parlor, finished in white and gold, opening on a terrace (on the Broad street front) covered in summer with an awning, and the billiard room, which is 40 feet square, with bays at the sides, with a large open fireplace of wrought stone and molded brick. In the basement are four bowling alleys.

Other clubs are the *Chatelet*, 1008 Broad street, and the *Progress* (Hebrew). There are also the *Newark Harmonic* and *Schubert Vocal* societies, and seventeen other singing societies.

ATHLETICS AND SPORT.—*Triton Boat Club:* Has a fine club and boat-house on the Passaic, north of Mount Pleasant Cemetery, with grounds which comprise a running track and tennis courts. It numbers about 100 members, and has a fine record. From the club-house is a straight-away course of 1½ miles, a view of which is commanded from the grand stand.

Other athletic and rowing clubs are the *Riverside Athletic Club*, whose grounds and house are near those of the *Triton*, the *Mystic, Eureka, Institute, Passaic* and *Atalantas*, the last formerly located on the Harlem, whence it is thought the increase in shipping, following upon the opening of the ship-canal, will drive most of the rowing clubs to Newark. The *Atalantas* are on the east bank of the Passaic, between Market and Centre street bridges.

Base-Ball Grounds, foot of Emmet street. Principal clubs: Newarks, Rosevilles, O. N. T.'s, Star Athletics and Tenth Ward Athletics.

Wheeling: The hard, level roads around Newark have made wheeling a popular sport. The Essex Bicycle Club is an important organization, and the New Jersey Wheelmen have their headquarters at 494 Broad street.

The New Jersey Trotting and Horse-Breeders' Association (Box 242, Newark, N. J.) is seeking to bring forward in the State an animal of superior characteristics for road and track purposes, and, by a careful system of breeding and training, to insure beauty, service, docility and speed.

The Mutual Driving Association: trots every Saturday afternoon, in summer, at Waverly.

TRANSPORTATION (see beginning of Chapter and title, Railroad Stations, p. 15). In the early part of this century the only public conveyances between Newark and New York were the stage-coaches. In 1800 there was but one of these, an ungainly vehicle with a long body, hung upon iron jacks, with five seats and a baggage

rack. Four miles of the road were so badly washed by the tide of the Passaic that the coach jolted over logs and stones, and the passengers usually preferred making that part of the journey on foot. In 1813, four lines from New York to Philadelphia passed through Newark. Of Gen. John Noble Cumming, who was one of the chief stage proprietors and mail contractors, an interesting anecdote is still preserved. During the period when Gideon Granger was Postmaster-General (1802–1809), there were serious irregularities in the delivery of the mails. Granger therefore determined to travel in disguise over the mail-routes. Some one at headquarters gave Cumming the tip, and he instructed his negro driver what to do should he happen to have a passenger answering a certain description. In due time the stage was boarded at Paulus Hook by Granger, and the negro driver with a crack of his whip sent the horses plunging over the rough road.

"Do you want to break all the bones in my body?" shouted Granger. "Drive slow! Drive slow!"

"Can't do it, sir!" was the reply. "I drive the United States mail!"

Again and again the Postmaster-General protested, but in vain; and by the time he reached Newark he was satisfied that on one route at least the interests of the United States mail service were well looked to.

It was in Newark that Roosevelt, in 1798, built the "Polacca," a little craft fitted with a steam engine of twenty-inch cylinder and two-feet stroke which, October 21, of that year, eight years before the successful trial of Fulton's "Clermont" on the Hudson river, made a trial trip on the Passaic, but with disputed success.

In 1818, a line of sloops and schooners was successfully established between Newark and New York. There was at one time much shipping in Newark, among the vessels clearing in 1837 being two whalers, the "John Wells" and the "Columbus." With the established success of steamboating this mode of travel was introduced between New York.

The first railroad trip between Newark and New York was made September 1, 1834, in the passenger car "Washington" which was, however, drawn by horses. December 2, 1835, the first engine passed over the road.

Now, Newark has railroad communication with the whole country.

The Newark and New York Railroad is a branch of the Central Railroad Co. of New Jersey. It was chartered March 1, 1866, and began operations in 1869. The branch between Newark, Elizabethport and Elizabeth was opened in 1872.

An idea of the active, progressive spirit which pervades Newark was given in the review of its industrial history (p. 2). The city is also an important trade center. The main business artery is Broad street, which has a width of 120 feet. This street was laid out by the original settlers. On it are large retail and wholesale stores of every description, Newark being the great shopping center for many towns and villages in Essex and Morris counties, enjoying also a large wholesale trade in the same localities. Market street is another important business thoroughfare.

The residential and business and manufacturing portions of Newark are beginning to be better defined. Of late years large residential districts have sprung up on the ridge which rises a little west of Broad street, and runs clear through the city north and south. High street from Market street to Clinton avenue is built up chiefly of elegant modern houses, those on the east side commanding from the rear an extended view of the valley of the Passaic and Hackensack. Mt. Pleasant avenue and its extension beyond Mt. Pleasant Cemetery form the eastern boundary line of another large and attractive residential district with many fine buildings, including the North End Club. On the plain which extends from the foot of the ridge to the Passaic are most of the numerous factories which have made Newark wealthy and famous, and the principal business streets. There are, however, on the low-ground, several spots distinguished for fine residences. These are Broad street, near Lincoln

Park, Park place, facing Military Park, and Washington place and street, facing Washington Park. The houses on these sites are mostly broad and spacious. On Park place is the Essex Club. These fine residences and numerous others of more modern proportions belong to people who have business as well as family interests in Newark, are thoroughly identified with the city's progress and are proud to hail from it. Herein Newark differs from those suburbs whose population is composed chiefly of New York business men. Within its boundaries many fortunes have been amassed, and it offers as attractive a field as ever for investment; business and professional men find ample employment; and there is steady demand for labor. Hence the vast majority of its residents are Newarkers in fact as well as in name. Newark is, however, also a pleasant dwelling-place for New York business men, because of its accessibility, its own attractiveness and the inexpensiveness of living there. Yet it offers all the conveniences of a city—electric light, gas, drainage, water-works, public buildings and institutions, churches, theatres, banks, insurance companies, newspapers and stores.

Newark has recently closed with the East Jersey Water Co. an important contract for the supply of pure water, which will add greatly to the advantages Newark has to offer as a place of residence. The Company has acquired water and water rights in the Pequannock water-shed and on the Pequannock river and its tributaries, located in the upper Passaic water-shed in the northern part of New Jersey. These have been acquired from the Lehigh Valley Railway Company, as lessee of the Morris Canal, and from the Society for Establishing Useful Manufactures and the Dundee Water Power and Land Co., its contract with the last two named giving it the perpetual right to divert surplus waters from the Passaic river or any of its tributaries.

CHAPTER II.

BERGEN NECK.

SUBURBS OF JERSEY CITY—The Newark branch of the Central Railroad of New Jersey cuts through Bergen Neck, with stations at Lafayette, Arlington avenue, Jackson avenue and West Bergen. These are parts of Jersey City which have been made so accessible to New York by the Central Railroad of New Jersey that people can reach their places of business in the city more quickly than those residents of New York itself who live above Fifty-ninth street, and many others living below that thoroughfare, who are off the line of the Elevated, or not near one of its stations. These suburbs are on the ridge which is virtually an extension of the Palisades, and from their high location they command superb views of New York Harbor and the city to the east and of Newark Bay and Newark to the west. The Statue of Liberty has become a familiar object to the dwellers on these heights, for it is visible from many points, and, though the tall lady turns her back on Jersey, the torch she grasps sheds its light on the shore of the stout little State which suffered so much in her cause. The bridge just beyond the Arlington avenue station is an excellent point from which to obtain the full rear view of the statue. From the western edge of the ridge, fine panoramas of Newark Bay, Newark and the Oranges burst into view. Their beauty is enhanced by the delicate green of numerous market gardens, covering the slope from the ridge to the bay, whose broad, glistening expanse shreds out into creeks and inlets which lose themselves among the sedges of the opposite shore. Over across the meadows are the factories whose smoke veils like a haze the busy city of Newark, so that its outlines become blurred and it looks like a great forest of houses, above which the church spires protrude as if they were the straight, tall monarchs of the woods, while in the distance the panorama is completed by the green plateau of the Orange Mountain.

After leaving Communipaw, the main stem of the Central Railroad of New Jersey continues along the shore, through what is also a part of Jersey City, stopping at Claremont, where there is a ground for trap shooting, and at Greenville, which enjoys the same advantages of situation and transportation to New York as those parts of Jersey City on the Newark branch. The next station is Pamrapo, which is part of Bayonne City.

BAYONNE CITY.—This flourishing community includes Pamrapo, Bayonne, Centreville and Bergen Point. Excepting at Centreville, where Constable Hook juts out with the enormous works of the Standard Oil Co. and the seaboard refineries of Lombard, Ayres & Co., Bayonne City consists of residential and business streets. While a city in its form of government and in the conveniences it affords, it wears the aspect of a charming suburb, most of the dwellings being surrounded with ample grounds, while its water-fronts on New York Harbor, the Kill von Kull and Newark Bay offer every facility for aquatic sports. From the harbor-front the yachtsman can speed down into the bay, up the rivers or around Staten Island—in fact, he has the choice of all the waters around New York. For rowing, canoeing and sailing in small boats, Newark Bay, which spreads out like a great lake, has long been a favorite sheet of water. The Argonautas and the New Jersey Athletic

Association have their boat-houses a little south of the Central's long bridge. Near by are the extensive base-ball and tennis grounds of the New Jersey Athletic Association, which enjoys a large membership, drawn not only from the immediate vicinity, but from Jersey City and the towns along the line of the Jersey Central. Trains stop for members of the club, and, on days when games are played, all trains stop at Avenue A station, near the grounds. This club adds greatly to the life of Bayonne City, and, with the amusements it affords the young people who participate in the sports and the older people who look on, is one of the chief attractions of this suburb.

Bergen Point proper has long been a favorite place of residence, so that, besides the modern houses, there are many old mansions, sheltered by ancient trees and with ample grounds—relics of the days when land was sold by the broad acre instead of by the front foot.

Historical Incidents.—The following historical incidents are of interest to all residents of Bergen Neck.

Communipaw, where the Newark branch leaves the main line, is the oldest settlement on the New Jersey shore of the Hudson. It is indebted for its name to Michael Pauw, Burgomaster of Amsterdam and Lord of Achtienhoven, near Utrecht, who obtained, July 12 and November 22, 1630, deeds from the Indians for a tract which he called Pavonia, a Latinization of his name. The shore had then already great commercial value because from there the Indians conveyed their peltry across to New Amsterdam. In the latter part of 1633 a house was erected at Communipaw, and this was the first actual settlement of which we have record. June 17, 1634, Jan Evertse Bout arrived in New Netherland with a commission from Pauw as his superintendent, and took possession of the house at Communipaw. He

BERGEN POINT.

varied the monotony of colonial life by falling in love with the daughter of a servitor, who returned his affection, thus causing such a "scandalum magnatum" (Bout being a family man), that the Schout of New Amsterdam crossed the river to remonstrate with him. But Bout flew into a passion, snapped his fingers in the Schout's face, and called him *een houd, een dief, een schobbejak* (a dog, a thief, a rascal), causing him to retreat, thus establishing at a very early date the principle of State sovereignty, for which New Jersey has always stoutly contended. July 20, 1638, Pauw having sold his purchase back to the New Amsterdam Company, Bout leased the "bouwerie" for six years for one-quarter of the crops, afterwards receiving, as a gift, a patent for the farm, the place being called in the deed Gamoenepaen.

In the early days of the Province, and especially under Keift's administration, troubles with the Indians were numerous and Communipaw suffered in common with the rest of Pavonia. In October, 1643, the Indians laid waste the four "bouweries" in Pavonia, including that of Bout at Gamoenepaen, destroying the buildings by stealthily creeping up to them through the bush and firing the roofs, which were of reeds or straw. So thorough were they in their work of devastation, that the country from Tappan to the Highlands of Navesink was once more in the hands of the aborigines. Peace was not concluded until the spring of 1645. Then Bout sold his farm to Michael Jansen and Claes Comptab.

In 1654 patents were issued for tracts "between Gemoenepaen and the Kil van Kol," most of them in what became afterwards known as Pembrepogh (Pamrapo). But the settlement of the country was soon again checked by another Indian uprising occasioned by a somewhat curious incident. Hendrick Van Dyck had planted with trees imported from Holland, a peach orchard on his farm in New Amsterdam, south of Trinity Church, between Broadway and North River. The Indians found the fruit so delicious that they made nightly raids. Finally, one night Van Dyck, who, gun in hand, had stationed himself in the orchard, fired upon the first dim figure he saw scaling the fence. His victim was an Indian girl. The Indians avenged her death chiefly upon the Pavonia settlements, and, excepting the family of Michael Jansen of Communipaw, not a man who did not seek refuge at New Amsterdam escaped with his life.

Owing to the devastation wrought during this uprising of the aborigines, Stuyvesant, then Governor, promulgated an enactment to the effect that all isolated settlers should remove before the middle of April, 1660, to the nearest village, so that the inhabitants could quickly unite for common defence. Communipaw was fortified in 1663, but meanwhile, in August, 1661, Stuyvesant granted permission for a settlement "behind Communepah" and "on a convenient spot which may be defended with ease." The settlement was begun on the hill and called Bergen. It was laid out in a square, the sides of which were 800 feet long. It was surrounded by a palisade, back of which was a street, and two streets crossed each other at right angles at the centre. Here was a public plot, and in its very centre a well was dug which remained in use during part of the present century, but was finally covered and a liberty-pole erected on the spot. This was taken down in the fall of 1870, and all traces of the well destroyed when the square was paved. The old land-mark might well have been retained, its removal being evidence of a callous disregard for historic tradition that is almost unintelligible.

Early in 1664 a log school-house was built. In 1790 the Columbia Academy was erected on the same lot, and gave way in 1857 to the present structure. In 1663 already the community had been taxed for church purposes, and with the following year the records of the Dutch Reformed Church, the first church in the State, commence. The original building was of logs, in what is now known as the old grave-yard. In 1680, an octagonal stone structure was put up. Around the wall were pews for the men, the women being accommodated with chairs on the floor. The bell was tolled from the centre of the church. This edifice was taken down and a new church put up in 1773. It stood until 1841, when the present structure was erected. Until 1809 the records were kept in Dutch.

The village passed quietly through the capture of New Amsterdam by the British, its recapture by the Dutch and subsequent cession to England ; and indeed, until the Revolution, nothing occured to ruffle its serenity. Even during the Revolution it was not subjected to much excitement. The Americans remained in possession of Bergen, which they had fortified, until Washington, who had made it his headquarters, decided to retreat to the Delaware. The British then moved in, and named the works Fort Delancey, in honor of the Westchester Tory, Oliver Delancey. It was garrisoned chiefly by Tory refugees, who made cowardly raids through Bergen Neck. Washington and Lafayette once dined under an apple tree near the Bergen Square. The tree was blown down during a great gale, September 3, 1821. When Lafayette, while visiting this country in 1824, passed through Bergen, he was presented with a cane made of wood from this tree. Bergen was consolidated with Jersey City in 1870.

In the latter part of 1660 the first road was laid out. It ran from Communipaw to Bergen—over the present Communipaw avenue to Palisade avenue, thence northerly along Palisade avenue to Academy street, thence westerly to the village.

While Fort Delancey was occupied by the Tory refugees, the people of Bergen were permitted to take provisions over the river to New York, where they would purchase household articles and other necessaries for themselves. They usually made the trip by way of Communipaw. The Tory refugees, scenting rich and easy prey, would waylay them on their way home and rob them. In order that they might be on their guard against these marauders the people arranged a code of signals, using therefor the half doors of a barn which stood just south of Communipaw avenue. If the upper half was open, it was a signal of safety ; if closed, it signified that the Tories were about. Then the people would remain on Ellis Island until the signal of safety was given. The Tories, learning that the doors were in some way used as signals, endeavored one day to lure some of the inhabitants ashore by means of false signaling. But, some of them maintaining that the open upper half of the door meant safety while others held as stoutly to the reverse opinion, the doors were opened and shut so rapidly that the people, suspecting something was wrong, remained on Ellis Island.

Communipaw came in for a share of Irving's pleasant raillery, he referring to it in his "Knickerbocker History" as follows :

"It is a well-known fact, which I can testify from my own experience, that on a clear, still summer evening, you may hear, from the Battery of New York, the obstreperous peals of broad-mouthed laughter of the Dutch negroes at Communipaw. * * * The negroes * * * carry on all the foreign trade, making frequent voyages to town in canoes loaded with oysters, buttermilk and cabbages.

"As to the honest burghers of Communipaw, like wise men and sound philosophers, they never look beyond their pipes, nor trouble their heads about any affairs out of their immediate neighborhood ; so that they live in profound and enviable ignorance of all the troubles, anxieties and revolutions of this distracted planet. They meet every Saturday afternoon, at the only tavern in the place, which bears as a sign, a square-headed likeness of the Prince of Orange, where they smoke a silent pipe, by way of promoting social conviviality, and invariab y drink a mug of cider to the success of Admiral Van Tromp, who they imagine is still sweeping the British Channel, with a broom at his mast-head. * * * The language likewise continues unadulterated by barbarous innovations ; and so critically correct is the village schoolmaster in his dialect, that his reading of a Low Dutch psalm has much the same effect on the nerves as the filing of a hand-saw."

August 24, 1780, Lafayette's light camp marched toward Bergen and the following morning appeared on the brow of the hill, east of the town, in full view of the enemy. The infantry foraged all the way down to Bergen Point, driving off cattle, loading wagons with grain and carting it away, leaving with the people certificates which might "procure for them, at some future day, compensation." There is, however, no evidence that these Revolutionary I. O. U.'s were ever honored. The British were much taunted for having allowed this foraging expedition to take place under their very eyes. A satire in verse, supposed to have been written by Susannah, daughter of Gov. Livingston, appeared soon afterwards. Some of the characteristic lines are :

We've almost, sweet sister, been frightened to death,
Nor have we as yet, quite recovered our breath.
An army of rebels came down t'other night,
Expecting, no doubt, that the British would fight.
Next morning we saw them parade at the Hook,
And thought to be sure this was too much to brook ;
That soon would the river be covered with boats,
With Hessians and English to cut all their throats ;

* * * * * * *

But this was all vision, Tabitha, to me,
Not an officer came, so much as to tea.
The Major himself, who has always some story
To lessen the worth of American glory,
Or ashamed to be seen or else of the day,
Would not venture to cross me though just in the way:
But stopped, like one shot at, then whisked up a lane ·
I'm sure the poor man felt a great deal of pain.
At length came the night, overloaded with fears,
And shew'd us on what we had leaned for five years.
The men who had wished for occasions for blows,
Now suffered themselves to be pulled by the nose.

A ferry was established between Bergen Point and Staten Island certainly as early as 1750, and probably before. The old route to Philadelphia was via Bergen Point and Blazing Star (Staten Island) ferries, and South Amboy. It took three days to reach Philadelphia, though the vehicle was called in the advertisements a "flying machine."

CHAPTER III.

ELIZABETH.

ELIZABETHPORT.—After leaving Bergen Point, the railroad crosses Newark Bay over a bridge nearly two miles long, with a pivot-drawbridge of iron spanning two openings of 75 feet each in the clear, and resting upon a pier of solid masonry. The first stop after Newark Bay has been crossed is Elizabethport, which occupies the water-front of Elizabeth. Here is an important junction for Newark on the north and for the Jersey Coast and Pine resorts on the south. Several rowing associations have their boat-houses at Elizabethport. There are numerous manufacturing establishments already here, and the favorable location of the place, together with the facilities for transportation, are constantly attracting others. Conspicuous, from the cars, is the large Singer sewing-machine factory.

ELIZABETH.—Public and Semi-Public Buildings and Institutions: *City Hall*, Elizabeth avenue, corner of West Scott place; *Court House*, Broad street; *Police Department*, City Hall (*supra*); *Post Office*, Arcade Building, Broad and West Grand streets; *Elizabeth General Hospital*, Jacques street, near Elizabeth avenue.

Railroad Stations: Central Railroad of New Jersey, Broad street, for Newark, New York, the Coast and Pine resorts, Lake Hopatcong, Budd's Lake and Schooley's Mountain; Philadelphia, Baltimore, Washington and the West, Spring street.

Elizabeth Public Library and Reading-Room, 21 Broad street, has 5,000 books for circulation and about 500 for reference; library and reading-rooms being free to any resident of Elizabeth, over twelve years of age, who can secure a proper guarantee to an application card. The circulation averages about 100 daily, and the card-holders number 2,035, as against 1,519 last year. The institution depends for its support upon voluntary subscriptions.

Few places have lately made such rapid progress as Elizabeth. For a number of years this city staggered under a heavy debt, and, at one time, an attempt was made to levy on the City Hall and other public buildings; but, now that the debt has been settled, the old town is enjoying an era of unprecedented prosperity. It is a charming place of residence, the rawness of modern development being softened by the mellow touch of historical association. Elizabeth is an old, even an ancient, settlement for this country, for it was the third spot in New Jersey to be settled, and the first by English-speaking people. When Gov. Philip Carteret made it, in 1665, the capital of the Province, there were only four log huts in the place. The Borough of Elizabeth Town was incorporated February 8, 1739; the City of Elizabeth was chartered March 4, 1863. Until comparatively recent years it was known as Elizabeth Town, and Elizabethport as Elizabethtown Point.

HISTORY.—Elizabeth was named in honor of the wife of Sir George Carteret. She was a friend of Pepys, who speaks of her in his diary: "She cries out of the vices of the Court, and how they are going to set up plays already. She do much cry out upon these things, and that which she believes will undo the whole nation."

The first white men to view the site of what is now Elizabeth, belonged to a little exploring party from Hudson's *Half Moon*, which September 3, 1609, had anchored in the Horse Shoe, Sandy Hook Bay. Sunday, September 6, John Coleman and four other men were sent to explore the harbor. They proceeded as far as Newark Bay. On their return trip Coleman was slain by an arrow shot by an Indian in concealment.

LAND TITLES.—Elizabeth was settled by Long Islanders who had previously emigrated from New England. The settlement was effected October 28, 1664, under a patent from Governor Nicolls, the settlers purchasing from the Indians, the final payment being made November 24, 1665. Carteret, who arrived in August, 1665, confirmed the rights by which the settlers held their lands. Nevertheless, this question of title became a serious matter of dissension between the townspeople, who upheld their rights under the Nicolls Patent and their purchase direct from the Indians, against the governors sent out from England by the Proprietors; and subsequently, when the Proprietors had ceded their rights to the Crown, even against those sent out under the royal authority itself. These contentions continued until the Revolution, and did much to develop in the people of Elizabeth that love of liberty, and above all of the right to manage their own affairs, independent of any proprietary or royal governors, which made them such staunch supporters of the Colonial cause during the Revolution. Indeed, the transition from the civil strife with the representatives of the Crown to armed resistance against the tyrannous edicts of the Crown itself was natural and easy. For, even during these civil dissensions, the spirit of the people had found vent in violent means of redress, such as tearing down the fences and destroying the property of those who had taken possession of land under deeds from the Proprietors; and, when the possessions of one of their sturdy republican fellows were confiscated, because of his opposition to the proprietary governor, his associates stood by him to a man, and raised among themselves a sufficient sum to indemnify him. Once these sturdy pioneers felt impelled by their love of local rights to unite in defending them, even with the hated representative of the Proprietors. When New York sought to extend its jurisdiction over neighboring provinces, the people of Elizabeth rallied around the proprietary governor and made common cause with him against the aggressor. For then it was province against province; but, with the first lull in the conflict, they showed themselves as ready as ever to combat the pretensions of the Proprietors and their representative. Such is an outline of the pre-Revolutionary history of Elizabeth and its vicinity.

The numerous details of the struggle would hardly be found interesting to the general reader. Those who desire to go more fully into the subject will find it treated of at length in Hatfield's History of Elizabeth.

The following scraps, which throw light upon life in Elizabeth before the Revolution, are taken at random from old records and newspapers :

May, 1666, a servant, "Robert Graij," runs away from Luke Watson, to whom he had bound himself for three years, and is advertised for as follows, the advertisement being entitled a "Hue and Cry" :

"His name Robert graij an Englishman bornd, about 20 yeares of age, a lustij bodied portely fellow. * * * It is supposed that he is in company with one Ruderic Powell, a pittifull fellow, who hath also absented himselfe and runn away."

A whaling company was organized February 15, 1668-9. This company captured the whale which came ashore in the cove on Sandy Hook, which since then has gone by the name of Spermacetti Cove.

October 13, 1679, Jaspar Daukers and Peter Sluyter, travelers from Friesland, visited Elizabeth. They lodged in a tavern at the Point kept by Frenchmen—"but there was nothing to be had there, except to warm us," and they lay down to sleep "upon a little hay before the fire."

"June 4, 1741.. Daniel Harrison Sent in his account of wood carted for Burning two Negros allowed Cury. 0. 11. 0."

"February 25, 174½. Joseph Heden acct. for wood to Burn the Negros Mr. Farrand paid allowed. . 0. 7. 0. Allowed to Isaac Lyon 4 / Curry. for a load of Wood to burn the first Negro. . 0. 4. 0."

"December 24, 1741. An Indian Wench named Sarah, absented some time from her Master the Rev. Mr. Simon Horton. * * * She is a short thick Wench about 24 Years of Age, and has lost some of her Fore Teeth."

"August 12, 1751. We hear from Elizabeth-Town, that two Women have been killed within these few Weeks past, near that Place, by falling out of riding Chairs."

"April 27, 1752. A lively Parcel of Negro Boys and Girls from 12 to 20 Years of Age, who have all had the

Small Pox, To be sold by Cornelius Hetfield, in Elizabeth Town."

"Sept. 19, 1763. Margaret Johnston (formerly the Widow Chetwood), who for many Years, kept the Nag's Head Tavern, near the Bridge in Elizabeth Town, begs leave to inform her old Customers and Friends that she now keeps a Public House near said Bridge in Elizabeth Town."

"Elizabeth Town (in New Jersey), January 23, 1764. Last Friday departed this life, Miss Mary Eldrington, an old virgin, in the 109th year of her age. She was of an ancient family, born at Eldrington-Hall, in Northumberland, Old England, and on the next day she was decently interred in St. John's Church-yard, at Elizabeth.Town.—It is remarkable, that, notwithstanding her great age, she was very desirous of getting a husband before she died; and not two years since, nothing could offend her so highly as to tell her that she was too old to be married." It will have been observed that this "old virgin" must have made a marvelously rapid journey from Eldrington Hall, in Northumberland, Old England, to Elizabeth Town, to be "decently interred in St. John's Church-yard," in the latter place the day after she was born in the former.

ELIZABETH DURING THE REVOLUTION.—The spirit which the Stamp Act aroused in the people of Elizabeth is shown by the following extract from a newspaper of the day:

"A large Gallows was erected in Elizabeth Town last week, with a Rope ready fixed thereto, and the Inhabitants there vow and declare that the first Person that either distributes or takes out a Stamped Paper shall be hung thereon without Judge or Jury."

Among the inhabitants were such sturdy patriots as William Livingston, the Revolutionary Governor of New Jersey; Elias Boudinot and William Peartree Smith. Commercial intercourse with Staten Island was broken off February 13. 1775, because its people had "manifested an unfriendly disposition toward the liberties of America."

During the night of July 4, 1776, but a few hours after the adoption of the Declaration of Independence, a British armed sloop, having run up on Elizabethtown Point, was attacked from the shore with two twelve-pounders, a great number of her men killed and the vessel set on fire and destroyed. This was probably the first military exploit of the new-born nation.

About the middle of December, 1776, the British made Elizabeth the base of operations against the militia camp at Chatham, but the New Jersey militia, under Col. Ford, met the enemy at Springfield, and, in an hour's attack, inflicted such damage that the British fell back under cover of night to Newark.

Shortly afte · Washington's successes at Trenton and Princeton, Gen. Maxwell, who commanded the militia, made a series of sorties from the Short Hills, had a successful skirmish with the Hessians at Springfield, January 5, 1777, and, about January 16, marched toward Elizabeth, and took possession of the town, making prisoners of detachments of Hessians and Highlanders and capturing valuable stores.

During the latter part of the winter of 1779, a plan was devised by the British to capture Gov. Livingston's and Maxwell's brigade, and an expedition under Lt.-Col. Stirling embarked for Elizabeth the night of February 24, 1780. The Governor, fortunately, was away, and Maxwell, having been apprised by a fugitive of the enemy's approach, had gathered his forces in the rear of the town. The British vented their chagrin by firing the Presbyterian Church and parsonage, the barracks and the Academy. Maxwell having determined, by the light of the burning buildings, the strength and disposition of the enemy's forces, fell upon them and drove them back to their boats, one of which grounded and was captured with all on board.

When the campaign of 1780 opened, Washington was encamped at Morristown, with posts thrown out as far as the Short Hills. The English organized an expedition against Washington's camp, with Elizabeth as a base. At day-break, June 7, the enemy, 6,000 strong, having landed from Staten Island, and led by Knyphausen, who was confident that the superior numbers and discipline of his troops would give him an easy victory, entered Elizabeth. An eye-witness describes the scene as one of the most beautiful he ever beheld. In the van rode a squadron of dragoons, of Simcoe's regiment, known as the "Queen's Rangers," with drawn swords and glittering helmets, mounted on very large and beautiful horses—then followed the infantry, composed of Hessians and English troops—the whole body amounting to nearly six thousand men, and every man, horseman and foot, clad in new uniforms, complete in panoply, and gorgeous with burnished brass and polished steel. The column proceeded along the Galloping Hill road, which leaves the Westfield road on the line of the Central Railroad of New Jersey, and, running northwest, enters Connecticut Farms, south of the Presbyterian Church.

On "Prospect Hill, in the rear of Springfield," the Americans kept a lookout, who, on being notified of the British advance, fired a signal with an 18-pounder and lighted a tar barrel. The militia hastened to their mustering places, and, by the time the British reached Connecticut Farms, had a sufficient force to oppose their progress. The vanguard, having been checked by a party of sixty militiamen, were driven back upon the main body. The Americans being afterward pushed back toward Springfield, a stand was made at the bridge over the Rahway, whence the enemy were repulsed so effectually that Knyphausen retreated under cover of night to Elizabeth, first having burned Connecticut Farms.

Thus the British expedition which had marched that morning in such gorgeous array and so confident of victory from Elizabeth was thrown back by a hastily-mustered, indifferently-equipped body of American militia. Before the British retired they laid waste the village of Connecticut Farms, one soldier brutally shooting down Mrs. Caldwell, the wife of Rev. James Caldwell, one of the most prominent Presbyterian clergymen in the country. Among the losses sustained by the British was that of Gen. Stirling, who received at the outset of the action a severe wound, from which he died a year later.

Piqued by the defeat of so gorgeous an array of horse and foot and flying artillery under Knyphausen, Sir Henry Clinton, June 23, took oversight in person of a second attempt to penetrate to Washington's camp from Elizabeth by way of Springfield and the Short Hills, with some 5,000 infantry, besides dragoons, and some 15 or 20 pieces of artillery—a force far superior to any which Washington could muster. Connecticut Farms was reached about sunrise, after the American pickets had been driven in. From this point the enemy proceeded in two columns, the right taking a somewhat more circuitous route on the north; the left the route leading direct from the "Farms" over the Rahway River to Springfield. Meanwhile the report of the 18-pound signal gun was reverberating through the Short Hills and the yeomen militia was hastening to the aid of Greene's Continentals.

Major Lee, with the horse and pickets under Capt. Walker, supported by Col. Ogden's command, was thrown forward to Little's Bridge on the Vauxhall road; Col. Dayton's regiment was entrusted with the defence of the village; Col. Angell, with his regiment and one piece of artillery, was posted at the bridge in front of the town, and Col. Shrieve at the second bridge to cover Angell's retreat. The enemy's left column spent the time which elapsed before the arrival of the right in manœuvers which led Gen. Greene to expect a flank movement. Hence, with his main body, he took position on the first range of hills back of Byram's tavern.

A portion of the enemy's right column having forded the stream, it was impossible for Lee with his inferior numbers to hold his ground. Angell and Shrieve were also forced back, but so obstinate was the stand made by the Americans, and so severe the losses sustained by the enemy, that the latter lost heart, and, anticipating a still more obstinate defence from the forces posted on the Short Hills, and learning that Washington had sent forward a brigade from Morristown, they fired the village, and beat a hasty retreat to Elizabeth pursued by detachments of militia, who picked off a red-coat wherever a chance offered, and so galled the flying British and Hessians that they crossed over to Staten Island, allowing the Americans to once more take possession of Elizabeth. On the American side not more than a thousand were engaged in this action.

While the fight at the bridge defended by Col Angell was at its hottest, the Americans found that their wads were giving out. Thereupon Chaplain Caldwell, whose wife had been so barbarously murdered at Connecticut Farms, as related above, rushed over to the church, and, returning with an armful of hymn-books, scattered them among the soldiers, shouting as he did so: "Now boys, give 'em Watts!"

Bret Harte has made this episode the subject of a highly dramatic poem, from which the following lines are quoted :

* * * They were left in the lurch
For the want of more wadding. He ran to the church,
Broke the door, stripped the pews, and dashed out in
the road
With his arms full of hymn-books, and threw down his
load
At their feet ! Then above all the shouting and shots
Rang his voice : "Put Watts into 'em—boys, give 'em
Watts!"

The conflicts at Springfield were small affairs even when compared with other battles of the Revolution, and, as far as numbers and casualties are concerned, they dwindle to insignificant skirmishes in comparison with the battles of modern warfare. Yet we may well feel proud that the wave of crimson and gold which twice swept over the plain from Elizabeth was twice broken by the little force of Continentals and militia which stood firm as a rock upon the hills at Springfield.

CHURCH HISTORY.—As in Newark, the oldest church in Elizabeth is the First Presbyterian. Mention is made as early as June, 1671, of a "town-house." This, like the "meeting-house" at Newark, doubtless served also as the Presbyterian place of worship. The lot included the present burying-ground of the First Presbyterian Church, and the town-house occupied part of the site of the present church. Graves were sometimes dug in the church, and, as the present building extends over what were portions of the burying-ground, the structure doubtless shelters the remains of several generations of the early settlers. No description of the first building remains. The church was burned down by the British the night of Tuesday, January 25, 1780. Services were held in a store-house until the autumn of 1785, when the new building was sufficiently near completion for occupancy. It was dedicated about January 1, 1786, and completed in 1791 or 1792.

Though the church never played so prominent a part in the political affairs of Elizabeth as the "Old First" of Newark did in those of that city, it is, historically, the most important structure in Elizabeth.

A curious figure in the early history of the church was Rev. John Harriman, who was called in September, 1687. Besides preaching, he ran a mill, a cider-press, was an agent for the sale of glass, dealt in real estate and negroes, was a surveyor and kept horses to let.

Rev. Jonathan Dickinson, who became pastor in September, 1709, became the first President of Princeton College, which was founded upon his classical school in Elizabeth, October 22, 1746 (*supra*, p. 10). His headstone may be seen in the Presbyterian burying-ground.

The Revolutionary pastor of the church was Rev. James Caldwell, whose wife was murdered by a British soldier at Connecticut Farms (p. 41, *supra*), and who supplied the Americans at Springfield with hymn-books for gun-wads, when, at a critical point in the battle, the latter were giving out. He was born at Cub Creek, Va., in April, 1734, graduated from the College of New Jersey (Princeton) in 1759, and became pastor at Elizabeth in 1762. In April, 1776, he was chaplain of Col. Dayton's regiment, which had been quartered in the town. At various times he acted also as Assistant-Commissary General. During the year 1778 he resided in Springfield, but in 1779 he removed to Connecticut Farms, so as to be nearer his people. He sometimes preached with his pistols lying on each side of him on the pulpit and sentinels on guard. He met his death in as tragic a manner as did his wife. On November 24, 1781, he was shot down, at Elizabeth Town Point, by an American soldier named Morgan. Caldwell had interfered in behalf of Miss Beulah Murray, who had arrived from New York, and whom the sentinel wished to detain

ELIZABETH.

until he had searched a bundle which she had tied up in a handkerchief. It is thought that the soldier had been bribed by the British to commit the act on the first favorable opportunity that offered. Morgan was hung for the crime, at Westfield, January 29, 1782.

Caldwell was buried beside his wife. A monument was erected over their remains, November 24, 1845, by a joint committee of the First Presbyterian Church and the Cincinnati of New Jersey.

The church stands on Broad street, between Murray street and Rahway avenue, forming, with the deeply-shaded grave-yard which spreads out in the rear, a restful break in the busiest thoroughfare of the city. In the rear wall are the headstones of two of Gov. Carteret's step-children, Samuel Lawrence (died August 16, 1687) and Thomas Lawrence (died October 26, 1687), the Governor's wife having been the widow of Capt. Wm. Lawrence, of Tew's Neck, L. I. These are the most ancient headstones in the cemetery. Between them, and, like them, of brown stone, is the Caldwell memorial. It might have been supposed that the virtues of Mrs. Caldwell and her tragic death would have preserved her memory from the exploits of the epitaph fiend. On the contrary, however, he seems to have had more license given him than usual, the following being the result of his mortuary throes:

Sacred to the memory of the Rev. James Caldwell and Hannah his wife, who fell victims to their country's cause in the years 1780 and 1781. He was the zealous and faithful pastor of the Presbyterian Church in this Town, where, by his evangelical labors in the Gospel vineyard, and his early attachment to the civil liberties of his country, he has left in the hearts of his people a better monument than brass or marble.

STOP, PASSENGER !

Here also lie the remains of a woman, who exhibited to the world a bright constellation of the female virtues.

On that memorable day, never to be forgotten, when a British foe invaded this fair village, and fired even the temple of the Deity, this peaceful daughter of Heaven retired to her hallowed apartment, imploring Heaven for the pardon of her enemies. In that sacred moment, she was, by the bloody hand of a British ruffian, dispatched, like her divine Redeemer, through a path of blood, to her long wished for native skies.

In addition to the First, there are six other Presbyterian churches. St. John's Episcopal Church, the oldest of that denomination in Elizabeth, is next in age to the First Presbyterian. It was founded by Rev. John Brooke, an Englishman and probably a graduate of Emanuel College, Cambridge, in 1700, obtaining his Master's degree in 1704. During the fall and winter 1705-1706, he preached at the house of a Col. Townley. Afterwards he was allowed to hold service in the Presbyterian Church, provided that he would not read the service of the Common Prayer Book. This prohibition he dodged by committing the service to memory. On St. John the Baptist's Day, he laid the foundation of a brick church—St. John's—which he describes as 50 feet long, 30 feet wide, and 21 feet high, with 9 windows. The history of the church is that of a natural and prosperous development.

Besides St. John's, there are three Protestant Episcopal churches.

Soon after the Revolution, the town was visited by traveling Methodist preachers. Asbury preached September 6, 1785, in the yet unfinished Presbyterian Church, and soon afterwards a society was formed. There are now five churches of this denomination.

There are 5 Baptist, 1 Congregational, 1 German Lutheran, 5 Roman Catholic churches, 1 Jewish synagogue, and 8 missions in Elizabeth.

MODERN ELIZABETH.—The historical incidents

related above have left traces in the Elizabeth of to-day. It wears the gentle aspect of mellow old age, which no new settlement can put on, no matter how vigorous an attempt it makes with Queen Anne and Colonial architecture, stained shingles, and andirons or spinning-wheels, acquired, like *Major General Stanley's* ancestors in the "Pirates of Penzance," by purchase. Elizabeth has the ripe color of genuine antiquity—but not the decaying look of senility. The old First Presbyterian Church structure, a relic of the last century, still re-echoes with vigorous preaching; the mansion of New Jersey's Revolutionary Governor, William Livingston, is occupied by his descendants, whose head is the grandson of Susan Livingston, who married a Kean and, on his death, the Polish patriot and author, Count Julian Ursin Niemcewitcz, after whom the manor was named Ursino. Yet with all these delightful historical associations there is an abundance of modern enterprise; while the ancient elms cast their soft shadows over the whole, causing the transition from the old to the new to seem less abrupt than elsewhere. The changes are taking place behind the veil of many traditions which, like the gauze curtains of a theatre, make the new scene seem less garish than if the lights shone full upon it. The elms are a grand feature of the city. Streets which might otherwise be uninteresting levels, are converted by them into colonnades of gray-ribbed columns, with arches of deep, dark green, in penetrating which the heat of summer loses much of its intensity. On either side are spacious mansions, in broad, well-kept grounds, while all around the old part of the city are new streets of residences.

Elizabeth offers all the conveniences of city life. Broad street, from the station to the First Presbyterian Church, is a fine business thoroughfare; and in the business and professional life of the place there is a felicitous mingling of old Elizabeth Town with new Elizabeth. The old families do not allow the new-comers to do all the "humping." They are active participants in the earnest, well-directed and successful efforts of the city to place itself abreast with the times. Public improvements are being made as rapidly as possible. The wooden pavements, which proved as disastrous to Elizabeth as the wooden horse did to ancient Troy (a large portion of the debt that acted as a clog on the city's progress was contracted for these pavements), are being replaced with stone, and soon, no doubt, Elizabeth will occupy the position to which she is entitled by virtue of her traditions as the old capital of the province, her facilities of transportation and her advantages of location, which should draw many manufacturing interests to her, and make the city as great a distributing center for her section of New Jersey as Newark is for hers.

NEW JERSEY JOCKEY CLUB.—One of the best mile tracks in New Jersey is that of the New Jersey Jockey Club, located within the corporate limits of the city of Elizabeth. It is about fifteen minutes' walk from the Union Depot, and about ten minutes from the Spring Street Station of the Central Railroad of New Jersey. The property of the Club, which includes 135 acres, is directly back of North Elizabeth and on the edge of the meadows fronting on Newark Bay. The track, already a good one, was built under the personal supervision of an expert, who also gave his attention to the erection of the Grand Stand, Club House, betting-shed, stables and other buildings. The stand is a substantial structure, with a seating capacity of about 3,000; it is a single floor, with restaurant, bar and toilet-rooms underneath. The elevation, however, is ample, not only giving a view of all that occurs on the track, but also across both the meadows and the Bay. The foundation

NEW JERSEY JOCKEY CLUB.

of the track is composed of peat and gravel, with a top dressing of sandy loam, procured at Great Island, about three-quarters of a mile distant from the track. The island is the property of the Central Railroad Company of New Jersey, who purchased it on account of the fine quality of the sand. It can hardly be excelled for track purposes by any similar pit in the State.

The Central Railroad Company of New Jersey has built a special branch directly to the rear of the stand, leaving the Elizabeth and Newark Branch near Great Island. The direct route is from the Liberty Street Ferry, by the Newark branch of the Central Railroad of New Jersey, and the trip is made every race day by half a dozen special trains of twelve and fifteen cars, in about twenty-five minutes.

Mr. Michael F. Dwyer (the younger of the Dwyer Brothers) is President of the Club, and the originator of the enterprise. Mr. H. D. McIntyre is the Secretary.

Racing on this fine track began October 16, 1889. It was at first intended to have two autumn meetings, to end about November 18th, and two spring meetings, to begin about April 15th and end May 14th. But after the middle of November, it is expected that there will be racing on three days each week—Mondays, Wednesdays and Fridays, as long as the weather permits.

WASHINGTON INAUGURATION CENTENNIAL. —An interesting event in our country's history was commemorated in Elizabeth when, April 29, 1889, President Harrison passed through the city on his way to the Washington Inauguration Centennial Celebration in New York. On the 23d of April, one hundred years before, Washington was received by the people of Elizabeth Town, partook of a repast at the residence of Elias Boudinot, and then drove to Elizabeth Town Point, where, after reviewing the New Jersey troops, he embarked for New York, his inauguration taking place April 30, 1789. President Harrison, Vice-President Morton, and their party, breakfasted at Gov. Green's house, and then reviewed a parade of militia, veterans, Freemasons, men attired as Revolutionary soldiers, and floats on which were scenes commemorating scenes peculiar to the life of our patriot fathers. The President and his party then proceeded to the point of embarkation for New York, passing under the "living arch," so called because its outlines were truly embellished with young girls chosen for their grace and beauty. These were dressed in white, carried banners emblematic of the States and Territories, and showered rosebuds and blossoms upon the President's carriage. Of all the features of the celebration, here and at New York, this "living arch," with the youth and beauty of the Elizabeth of to-day adorning a memorial of the Elizabeth Town of our ancestors, is said to have most charmed the President.

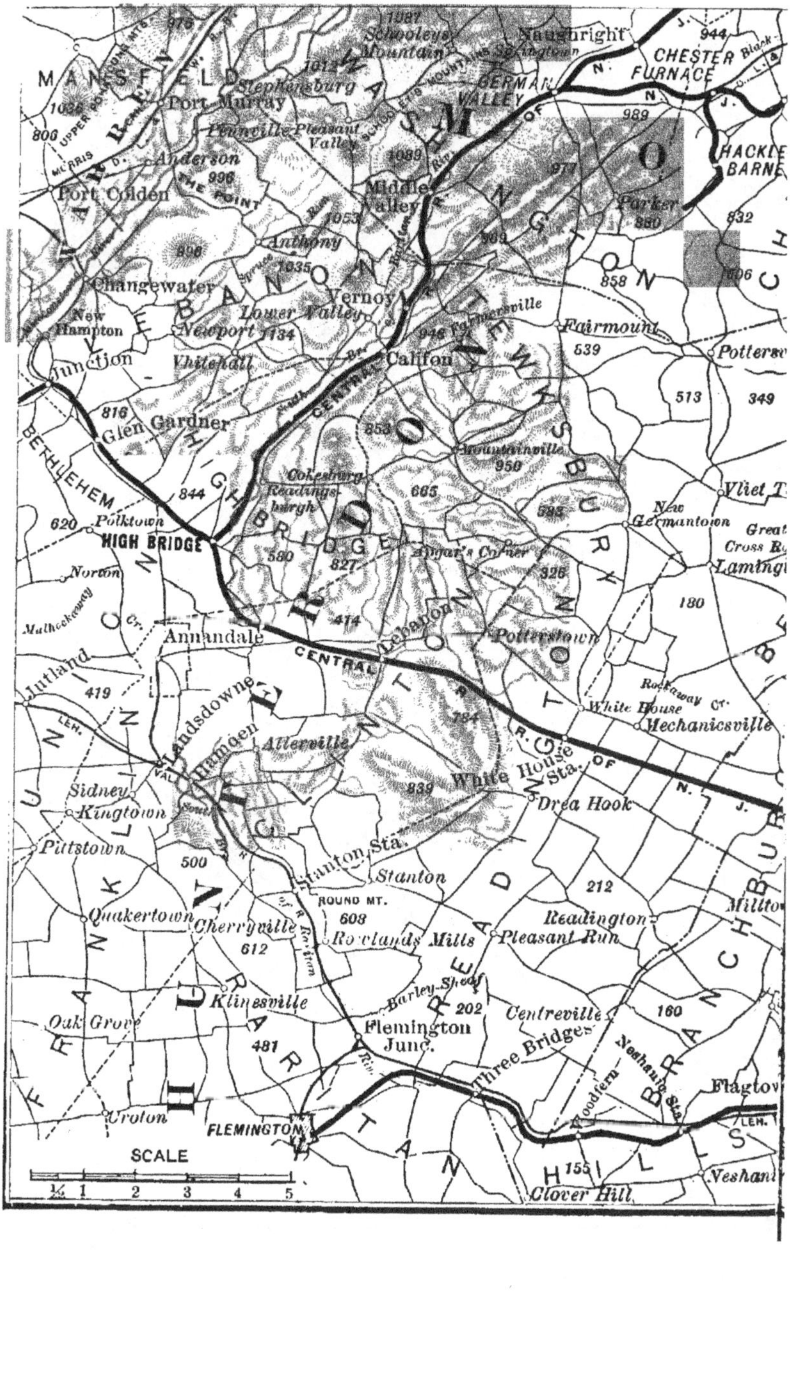
Schooleys Mountain
Naughright
CHESTER FURNACE
MANSFIELD
Stephensburg
GERMAN VALLEY
Port Murray
Pleasant Valley
Anderson
Port Colden
Middle Valley
HACKLE BARNE
Parker
Anthony
Changewater
Vernoy
Lower Valley
New Hampton
Newport
Farmersville
Fairmount
Junction
Whitehall
Califon
Potters
Glen Gardner
BETHLEHEM
Mountainville
Cokesburg
Readingsburgh
Vliet
New Germantown
Polktown
HIGH BRIDGE
Great Cross Rd.
Lamingt
Norton
Potterstown
Annandale
CENTRAL
Lebanon
Jutland
Rockaway Cr.
White House
Mechanicsville
Landsdowne
Hamden
Allerville
Sidney
Kingtown
White House Sta.
Drea Hook
Pittstown
Stanton Sta.
Stanton
ROUND MT.
Quakertown
Cherryville
Readington
Rowlands Mills
Pleasant Run
Millto
Barley Sheaf
Klinesville
Centreville
Oak Grove
Flemington Junc.
Three Bridges
Neshanic Sta.
Woodfern
Flagtov
Croton
FLEMINGTON
Clover Hill
Neshani
SCALE
½ 1 2 3 4 5

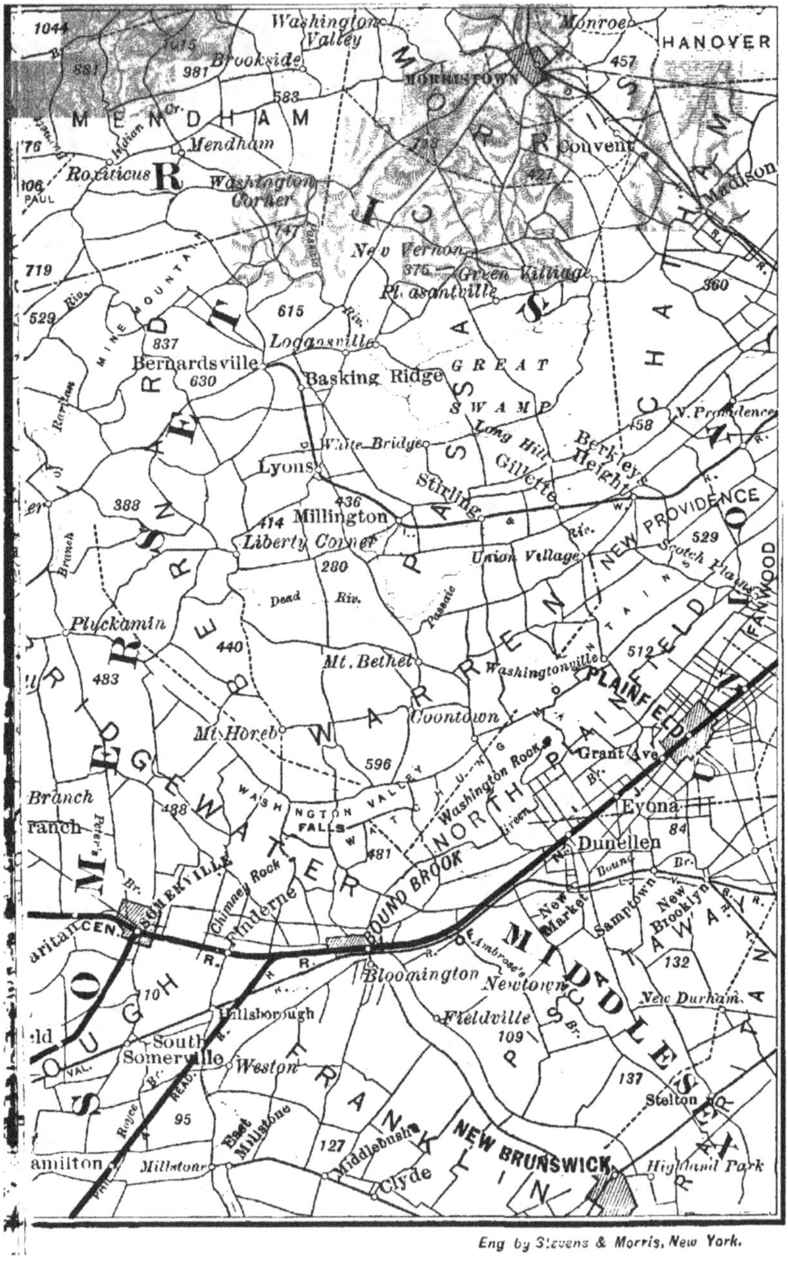

Eng by Stevens & Morris, New York.

CHAPTER IV.

From the Kills to the Delaware.

At Elizabethport, the main stem of the Central Railroad of New Jersey leaves salt water and crosses the State through fertile plains and picturesque mountain passes to the Delaware.

After a stop at El Mora, a suburb of Elizabeth, Roselle, the first of a long line of attractive suburban settlements, is reached.

ROSELLE.—Roselle has no history, but if it cannot boast an interesting past it has a charming present. It is a lovely little village with broad level streets, the trees spreading their boughs so as to form a leafy arch overhead. The grounds around the station are prettily laid out, and at the head of the grass plot on the south side of the track is a tall flag-staff, which has the honor of being marked on the maps of the State Geological Survey. Roselle enjoys the advantage of the local rapid transit service on the Newark and Elizabeth branch of the Central Railroad of New Jersey, in addition to the New York trains.

About a mile south of Roselle is Wheat Sheaf, a cluster of houses which includes the old Wheat Sheaf Inn. It stood in Revolutionary days, and Washington and Lafayette are said to have dined there.

CRANFORD.—Like Roselle, Cranford has developed from a farming settlement into a place of suburban residence. Since building operations have been under way on a concerted plan, the place has been steadily growing. A portion of the village, with lots sloping down to the

Rahway River, has been drained and sewered, and there houses are let as fast as they can be built. The Rahway River, which winds its way in the deep shade of the trees which fairly line its banks, has been dammed by the Cranford River Association and been made navigable for some three miles, and there are now about eighty row-boats and canoes and one steam-launch on the river. Regattas and boat parades are among the enjoyments at the disposal of Cranford residents. Of a moonlight night the Rahway River is a most romantic stream. The moonbeams, piercing the foliage of the leafy arch, are shredded out over the placid water, while now and then a star is seen peeping through some break in the foliage.

WESTFIELD.—This is an old settlement. It dates from 1720, and the Presbyterian Church—whose graceful white spire, piercing the foliage in which Westfield is embowered, is a prominent feature of the views to be had from the Watchung Mountain—from 1730. Life was quite primitive in those early days. The men made plows with wooden shares; pretty much every man was his own tanner, and the women spun all the clothing. A singing-teacher advertised singing-lessons at $1.00 for thirty nights or $2.00 a quarter, "subscribers to find their own wood and candles";—the Squallinis and the Katzenheulers might have found Americans in those days as willing, but hardly as lucrative victims as now.

Early in this century, life in Westfield was rendered more joyful by the "Stage House," which was noted for its flip. This pre-Raphaelite beverage, besides gladdening the soul of Westfield's population, gave the house a reputation among the *haute volée* of New York and Philadelphia who traveled this way. As the stage rattled up to the hostelry of a winter morning, the door of the house was thrown open, emitting the crackle and sheen of the logs that blazed on the huge open hearth, while in

the door-way stood the host, old Charley Gilman, the brass buttons on whose blue coat were not more shiny than his rubicund face. When the benumbed passengers had gathered around the hearth, a quart jug for each was nearly filled with malt-beer, sweetened, and brought to a foam with a red-hot poker. With a half-pint of rum poured in and nutmeg grated on the foam, the flip was ready.

Westfield to-day is a thriving, progressive place. Broad street is a bustling business thoroughfare, with shops of all kinds. The older streets are spanned by the boughs of grand old trees, and the spacious grounds which surround the ample dwellings are cool and shady. The newer streets have rows of pretty, modern houses. The ground here begins to be rolling, for it is nearer the foot of Watchung Mountain than the places already spoken of. There are slight elevations on either side of the railroad, and from the southerly height one has a pretty view over the town, almost hidden among the trees, along Watchung Mountain and across the broad break in this ridge of trap at Millburn, to the softly wooded slope of its continuation at Wyoming. The surrounding country is rolling and picturesque—sweeps of meadow-land broken up by thickets and clumps of trees, the greens ranging from the light verdure of grasses and bushes to the rich, deep tones of the woods. A similar view of Watchung and the lands along its base, is had from the rise of ground north of the railroad on which the old church stands. Here, too, is the ancient grave-yard with its many venerable, lichen-specked headstones.

A little more than two miles northeast of Westfield is the hamlet of Branch Mills, where a pretty lake, Echo, has been formed by damming Normahiggin (Clear Stream) Brook. There is good bass-fishing in the lake. Along the mill-stream was an Indian burying-ground.

On Watchung Mountain, two miles north of Westfield, is Baltus Roll, a clearing with a few cottages, named after an old man who many years ago was murdered there in his hut. From this clearing one has a view across the plain to New York Bay; the Statue of Liberty, the Bridge, and other prominent points being easily discernible. The panorama is almost as grand as that from Washington Rock (p. 58).

FANWOOD.—Here is the station for SCOTCH PLAINS, a settlement founded by Scots in 1684. The Baptist Church, dating from 1742, was the parent of the first Baptist Church in New York. The pretty little village nestles in the bosom of Watchung. Fanwood itself, is a park-like place of residences among beautifully laid out grounds and has an air of elegance and refinement. It boasts a fine club-house with bowling-alleys, billiard-room, etc. Taken all in all, Fanwood is a little gem.

In the valley beyond the notch at Scotch Plains, is Feltville, once romantically known as the "Deserted Village," because, when the factory there ceased operations, it and the dwellings about it were vacated. It is now a pretty place of residence called Glenside. Fanwood is the nearest station to it. Feltville Lake is a very pretty sheet of water.

NETHERWOOD.—This is also a park, with a number of attractive residences and several large villas. In addition, however, it has the most spacious and best kept hotel in Central New Jersey—the "Netherwood," a large brick building, with tiled floors, rooms of ample dimensions, and a broad, lofty piazza, which affords a pleasant promenade in the cool of the evening. Being within forty-five minutes of New York, it enjoys an unusually long season, and is a spring and fall, as well as a summer resort for New York business men and their families, generally opening about May 1st, and not clos-

FRENCH'S POND
FRENCH'S POND ABOVE WETUMPKA FALLS PLAINFIELD
OLD HOMESTEAD. AT DUNELLEN
L. FERAUD
FROM THE NETHERWOOD

ing until October 1st or even well into that month, according to the state of the weather. It is a clean, roomy, comfortable hotel, with a plain but excellent table, and for its rates ($2.00–$3.00 a day for transients, with reductions for those remaining a week or longer) furnishing extraordinarily good accommodations. The "Netherwood" is one of the few hotels near New York open during the spring and fall, and, as we are beginning to appreciate more and more the charms of country life during blossom-month, when all Nature is in an ecstatic tremor ; and also during September and October, when Autumn flings a mantle of yellow and crimson over mountain and plain—and on no mountain is she more lavish with her colors than on Watchung, which glows all day long like a slice of sunset cut from the previous evening's sky—Netherwood should become as much of a resort in spring and autumn as it is in summer. Indeed, it is a question if, with a toboggan slide and an artificial sheet of ice for skating, it could not be made a winter resort, especially as the mountain must somewhat protect the place from the blasts of the northwest winds.

DRIVES.—The roads for miles around Netherwood are macadamized, and kept in excellent condition. A favorite drive is over the Johnston road, which ascends Watchung through the notch at Scotch Plains and then proceeds over the mountain, affording glimpses of the grand expanse of plain through the foliage until it descends into the pass back of Plainfleld, a short distance below Wetumpka Falls. These falls are well worth a visit, as they dash picturesquely over a rocky declivity, especially after a heavy rain. They somewhat resemble Buttermilk Falls, near Bound Brook (see illustration, p. 63). Driving along Stony Brook toward Plainfield, there is another pretty fall formed by a mill-dam (see illustration, p. 57). Instead of taking the Johnston road one can continue through the notch to Feltville or Glenside (p. 54). The return from Glenside can be by way of Baltus Roll (p. 54). Often the drive to Glenside is extended on to Summit, and the return made through Springfield (p. 41), a route of about twenty miles. To Rahway, seven miles, and to New Brunswick, ten miles, are also pleasant drives. A picturesque way of reaching New Brunswick, is to drive to Bound Brook, and then along the Raritan, returning the usual way (twenty-five miles). A very pretty circuit can be made by way of New Brooklyn and New Market ponds, through Dunellen to Washington Rock (pp. 58, 61), continuing north and east through Washingtonville, in the pass back of Plainfield, which is reached by way of Wetumpka Falls (fourteen miles). This drive enables one to see the chief points of interest in this district. It should be taken on a clear afternoon, in order that the wonderful panorama from Washington Rock may be seen at its best (pp. 58 and 61, and illustration p. 61). Other points of interest are Chimney Rock and Buttermilk Falls, near Bound Brook (see pp. 62 and 63).

Carriage Hire.—Single team $1.00, double team $2.00 per hour; saddle horse, $3.00 for morning or afternoon; board, $30.00 a month.

PLAINFIELD.—Plainfield is one of the vigorously progressive places in New Jersey. It seems to enjoy continued prosperity. In 1860 it had a population of 3,224 ; its population now numbers 13,000. Since 1870 its valuation has risen from $3,102,295 to $5,712,115.

It lies in the corner of three counties. Plainfield proper is part of Union, North Plainfield of Somerset and South Plainfield of Middlesex. North Plainfield is locally called "the Borough," but, practically, it and

OLD FRIENDS' MEETING-HOUSE.

Plainfield form one place. Every resident of Plainfield knows, however, that politically they are separate. For, when the city "votes dry," the borough offers a place of refuge to those socially inclined, who can, from the northern bank of Green Brook, bid defiance to the reformers on the southern bank. The brook, though a narrow dividing-line topographically, is a very broad one politically.

Plainfield boasts a number of large manufacturing establishments, including machine-tool works, printing-press, clothing, air-pump and oil-cloth factories, foundries and grist-mills; and it has several streets devoted to business of every variety. In its factories and stores it is a city, not differing in its aspect from other flourishing cities of the same population; but over the plain to the south is a broad band of residential streets, and toward Watchung on the north are many fine country places.

Plainfield probably owes its prosperity equally to its business enterprise and its charms as a place of residence. To begin with, it is a very healthful spot. The air is decidedly beneficial in pulmonary complaints. One of the highest New York authorities on these diseases has advised patients to go either to Colorado or Plainfield, speaking of Plainfield as "the Colorado of the East." It may be said to drain itself, for its soil is sand and gravel, so that this whole district can be likened to a huge patent filter. The roads are dry almost before it has stopped raining. There seems to be a steady flow of water under Plainfield from north to south. The matter was tested by pouring kerosene into a well in North Plainfield, and in due time the taste of the oil was communicated to several wells in the southerly part of the city. The residential part of Plainfield south of the railroad (which, by the way.

is elevated) is beautifully laid out. The streets run in broad avenues, shaded by superb trees, with lines of fine residences and ample, well-kept grounds on either side. Closely-trimmed lawns, flower-beds and shrubs vary the level expanse. This part of the city seems to have been developed as a whole, and with the one purpose of making it unsurpassed for spaciousness and beauty, a result which has certainly been attained. On the mountain side, north of the railroad, are a large number of handsome residences and extensive country places, that of Mr. John Taylor Johnston, formerly President of the Central Railroad Company of New Jersey, being considered one of the finest.

Some of the grandest sites for residences, within a reasonable distance of New York, are on Watchung Mountain; but this fact does not seem to have been appreciated here as it has been at Orange, for instance, where along the top of the Orange Mountain (a continuation of Watchung) is a series of beautiful country places. Watchung Mountain is not broken up in peaks. Its top is a broad plateau. Wherever there is a break in the woods along its southern edge, as at Baltus Roll, near Westfield, at points on the Johnston drive, near Plainfield, and at Washington Rock, near Dunellen, one's vision is fairly astounded by the superb panorama, which gives a sweeping view across the plain to the Highlands of Navesink, Staten Island, New York Harbor, with the Statue of Liberty, Brooklyn Bridge and other prominent points, while in the foreground and middle distance are the tree-embowered towns and villages of the level land, which stretches out at one's feet for a distance of some twenty miles. By thinning out the trees a little on the southerly edge and slope of Watchung, one could have a series of residence sites extending from Baltus Roll to Bound Brook, for twelve miles, each of which would command a view equal to that to be had from Washington Rock—a view famous all the country round.

Plainfield offers other attractions as a place of residence besides location and careful topographical development. Socially and intellectually it is a prominent locality. It has in the Job Male Library and Art Gallery, Park avenue and Eighth street, a public institution which cannot fail to stimulate intellectual and artistic tastes. Here, besides a fine collection of books and of paintings by well-known foreign and native artists, is the Schoonmaker collection of porcelains and cloisonné, the rare pieces of the latter outnumbering those to be seen in any museum in Europe or America. In few places outside of the great cities can one have, as in Plainfield, a Sang-de-bœuf crackle vase for a near neighbor, or be on a footing of intimacy with a cloisonné *jardinière*.

Besides a number of excellent private schools, Plainfield has public schools which have elicited warm praise from the State authorities. The buildings are named the Washington, Franklin, Irving, Stillman and Bryant; the last, on East Sixth street, near Richmond, being well known and much admired in educational circles. The Chautauqua University also has its headquarters at Plainfield.

There is an unusually good base-ball club, the Crescents, a large bicycle club, a flourishing bowling club, an amateur photographic association, whose members turn out excellent work, lawn tennis and athletic clubs, and a driving association with a driving park at Avon. There is quite a large German element in Plainfield, and the list of clubs would be incomplete without mention of the Turners, Saengers and Schuetzen.

Plainfield has 4 Presbyterian, 4 Episcopal, 3 Baptist, 2 Congregational, 1 Seventh-Day Baptist, 1 Methodist-

Episcopal, 1 Hicksite, 1 Friends and 2 Roman Catholic churches; and a Bethel Mission for colored people. The Young Men's Christian Association also has a flourishing branch. Muhlenberg Hospital is on Fourth street.

There are fifty miles of macadamized roads around Plainfield. Drives and rates of carriage hire will be found under Netherwood (p. 56).

Plainfield has such a vigorous, interesting present that it is difficult to realize its existence in any previous undeveloped state; and, indeed, though its history runs back over two centuries, there are few facts of note to chronicle. It was settled in 1684, the pioneers being Thomas Gordon, John Forbes, John Barclay and Robert Fullerton. The first frame house was put up in 1735; the first school-house (Front and Peace streets) in 1760; the first grist-mill in the same year on the upper mill-pond, being removed in 1790 to Somerset street, where, on the north bank of Green Brook, a barn is pointed out as the old building; the Friends' Meeting-House, which still stands in a shady nook on Peace street, in 1788; the first factory (hat) in 1808, when a post office was also established. Plainfield Township was created in 1846, and Plainfield became a city in 1869.

DUNELLEN.—Grant avenue and Evona are suburbs of Plainfield. Dunellen is named after Edward Dunham, of New Market, about a mile south of the railroad. About the year 1700, Dunham reproved some one for laboring on Sunday. Asked for his authority, he searched the Scriptures and became convinced that Sunday was not the Bible Sabbath. He founded a Seventh-Day Baptist congregation, which still exists, their first church having been built in 1736. On Saturday the church bell summons the worshippers, and business in New Market is suspended until Sunday, when it is resumed with vigor. New Market is a pleasant annex to Dunellen. There is

Bryant Public School,
Plainfield, N. J

Oscar S. Teale. Architect.

good fishing in the little pond (Spring Lake), which is connected with that at New Brooklyn by a stream deep enough for row-boats.

There are at Dunellen a number of pretty residences. On the grounds of one of these is a genuine English bowling-green. Two interesting features of the place are Washington Rock and the ancient-looking Runyon House, which is passed on the way to the Rock. Washington Rock is an immense trap boulder on the southerly edge of Watchung, commanding a superb panoramic view across the plain to the coast, which is described, p. 58.

In the spring of 1777, and the winter of '78–'79, when Washington's army was at Bound Brook and the enemy at New Brunswick, a look-out was kept from the Rock, which commands a view of New Brunswick and the intervening country, so that Washington could be kept informed of every movement made by the British. He himself surveyed the country from this rock, a circumstance to which, of course, it owes its name.

BOUND BROOK.—So many trains pull in and out of Bound Brook daily as to make it an important railroad center. From here the fast Philadelphia, Baltimore, and Washington express trains dash southward over the Philadelphia and Reading track, and, on their return, rush toward New York at a rate of speed which brings Bound Brook within forty-five minutes of the city. With its railroads, the Raritan Canal, and the Raritan River, it is small wonder that the town has taken on new life.

It is delightfully situated, the river offering facilities for boating and fishing, while but a mile from the station is our old friend Watchung, with an inviting slope for residences. The growth of the place within the last decade has been conservative but steady, among those settling there being men of substantial means who have identified themselves with the place by aiding in developing its industrial resources; not simply residing there and going to New York to business, but establishing manufactories, so that there are now at Bound Brook woolen mills, brass works, and pump, paint, and car-heater and ventilator factories. There are water-works, and, though the place drains naturally to the river, several streets have already been sewered and the system is to be extended. Like Rip Van Winkle, the town has at last awakened, but, unlike Rip, appears rejuvenated after its long sleep.

A prominent feature of the landscape around Bound Brook is a trap boulder on the east side of the entrance to the notch which leads into Washington Valley. From its peculiar shape it is called Chimney Rock, and, as it is painted white, it is conspicuous from many points. One can drive to the little cluster of houses and to the excursion grounds at the foot of the mountain, about one and a half miles north from Bound Brook, but, to reach the rock itself, it is necessary to ascend a steep and stony path. The view, however, up toward the valley, well repays the climb. The road curves around the foot of Round Top on the opposite side, and then disappears in the gorge whose winding course is marked out by the furrow in the soft green with which the mountain slopes on either side are bedecked. In the distance beyond this verdure are the blue hills of Morris County.

After descending from Chimney Rock, it is worth while to ascend Round Top. For from here one can look up the gorge and see Middle Brook dashing over the boulders at Buttermilk Falls and then rushing over its rocky bed through the notch into the plain, where it meanders tranquilly through the rich meadowland until its waters, having swelled the stream of Bound Brook, flow into the Raritan.

From the path at the foot of Chimney Rock a trail leads to Buttermilk Falls. Here, when it has been raining sufficiently to cause a free overflow from the reservoir above, the water plunges over a wall of solid trap into a rocky basin, dashes in and out among boulders, frothing and foaming, as though with anger that anything should dare to check its course, and then rushes on with wild abandon through the gorge. With the wall of trap the sides of the basin form a horse-shoe, seemingly hewn out of solid rock by some titanic graver of pre-historic days.

Revolutionary Memories.—Bound Brook is rich in Revolutionary memories. After Washington's masterful strategy in crossing the Delaware and turning a retreat into victory at Princeton and Trenton, he marched, January 4, 1777, northward, crossed the Raritan and encamped at Somerset Court House (now Millstone), resuming his march the next day to Pluckamin, where he halted two days. There died Capt. Leslie, a British officer, mortally wounded at Princeton. A plain monument marks his grave. From Pluckamin the army went into winter quarters at Morristown.

The following spring (May 28th) Washington moved his headquarters to the "Heights at Middlebrook"—Bound Brook—in order to be within striking distance of the enemy at New Brunswick. A force under Gen. Lincoln had already been stationed there in April. His headquarters were in the only two-story house the village could boast. The location of Washington's forces was on the right of the road through the gorge in which Chimney Rock is situated, just where it rises up from the bed of the little stream to the level of Washington's Valley. All approaches were strongly guarded. Huts for the officers were built east of the rock at the edge of the woods. A redoubt commanding the bridge north of the railroad crossing was thrown up at Bound Brook, to check an attack from the direction of New Brunswick. On the apex of the Round Top, on the left of the Chimney Rock gorge, was a rude hut which Washington frequented during those anxious months, while on the east side of the gorge and fronting the plain, was Washington's Rock, both good points of lookout. June 13th the British moved from New Brunswick to Somerset Court House, and lay there till the 19th, when, finding that Washington could not be lured from his almost impregnable position, they returned to New Brunswick, which was evacuated June 22d, the enemy retreating to Amboy. Washington then moved down to Quibbletown (New Market). Howe, learning of this, made a sudden retrograde movement; but, being met at every cross-road by small bodies of militia, who made all the resistance in their power, his manœuvers were checked and delayed until Washington became informed of them, and moved back to his former strong position. The British again turned on their weary march toward Amboy, harassed flank and rear by Scott's Light Horse and Morgan's Rangers.

In December, the year following, a large portion of Washington's army went into winter quarters at Bound Brook. The artillery, under Knox, was posted in earthworks the remains of which may still be seen near Martinsville, in Washington Valley; Washington and his wife occupied, until June, 1779, the Caleb Miller house at Somerville, where it may still be seen; and Baron Steuben's headquarters, the Abraham Staats house at Bloomington, near Bound Brook, is still standing. February 18th Knox gave a ball, supper, and a display of fireworks at Pluckamin, in honor of the anniversary of the alliance with France. Late in April a review and parade were held in honor of Mr. Gerard, the French minister, and Don Juan de Miralles, an unofficial agent of Spain. After the military display Washington, his generals and the foreign guests were entertained by Steuben at dinner, the table being spread under the trees in front of the baron's headquarters.*

The Staats house, where Steuben had his headquarters, is one of the most interesting relics of the Revolution. For, as it has remained in the possession of descendents of the family, it still presents much the same appearance as in 1778–'79. Though it has been added to, the exterior of the old portion is entirely unchanged, and the interior is filled from parlor to garret with old furniture and china, which the family have taken great pride in preserving. The house, a long, rambling

* See "The Story of an Old Farm," by A. D. Mellick, Jr.

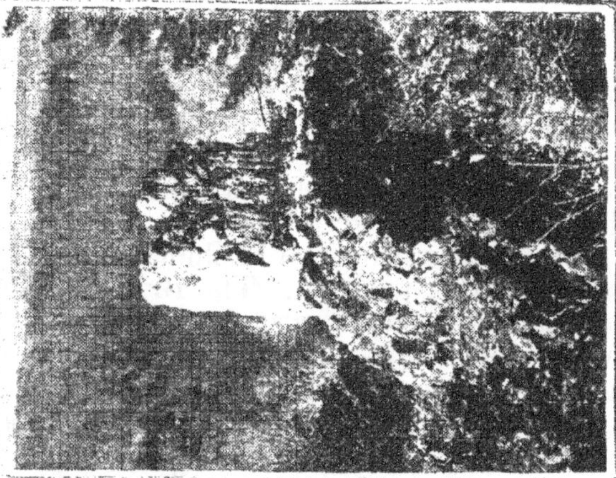

white building, stands on what is now the Latourette farm, on the south bank of the Raritan. A lane leads from the road to the grounds, shaded by trees under which, in the spring of 1779, Steuben entertained Washington and his Generals at dinner. The present parlor is the old dining-room. Here are chairs which were in the house during the Revolution, and china and glassware in a deep, quaintly arched closet with doors shaped like church windows—among the glasses being one which Washington drank from, while the whole table service, table and chairs must have been used by Steuben and his guests, one of whom doubtless Washington often was. There are also several pieces of old china on the mantle over the fire-place, among them statuettes of Minerva and Milton, which, with many other articles of fine ware, were buried near the house by a New Brunswick merchant, who feared the British would loot his store and who, when he dug up his stock, presented these statuettes to the Staats. The portraits of the old couple—kind-looking souls—hang on the parlor wall. While a modern range has been placed in the kitchen, the grand old fire-place has not been bricked up, and there still hangs the old crane, huge enough to serve as a derrick. On the landing of one of the six flights of stairs, leading from the ground floor to the second story, is an old clock. Many of the bedsteads have high posterns. Among the furniture is a chair which, it is claimed, was in the *Mayflower*. Five generations of the family have been born in this house.

Another interesting building is the old Fisher tavern at Middle Brook, which is virtually part of Bound Brook. It stands at the railroad crossing just south of the tracks. It was a tavern during the Revolution and is still one, being kept, and well too, by a grand-niece of the Revolutionary proprietor. It is a low, dun-colored house, directly on the street, with a quaintly slatted well at its western end. The low ceiling of the tap-room is indented by bayonet points and gun muzzles; and in the room above the first Masonic lodge in the United States was organized.

On the road to Finderne, on an elevation back of the mill, north of the railroad crossing, there stands among tall pines a long, low, ramshackle structure, known as the Van Horn house. During the Revolution this was occupied by the Herberts, who rejoiced in five pretty and engaging daughters. Old Herbert kept open house for friend and foe alike, his hospitality to British officers at times bringing down upon him the contempt of his patriotic neighbors. But, by the end of the war, his daughters were, between officers of the two armies, not only engaging but engaged, a state of affairs as satisfactory to the hospitable father as to the young ladies themselves. Whether the marriages were three to two in favor of the American forces, or whether the enemy carried off the larger share of the fair booty, history unfortunately does not record. For Washington's headquarters, see pp. 67, 68.

SOMERVILLE.—At Finderne, a station between Bound Brook and Somerville, are several large stock-farms, the country throughout this section affording excellent pasturage. During Washington's winter encampment General and Mrs. Greene resided at the Van Veghten house, a brick structure erected early in the last century, on the Raritan, a short distance southwest of the Finderne railway station. A brigade was encamped near by. Major Henry Lee, "Light Horse Harry," was quartered at the Herbert house ; and it was probably during this encampment he first met Mrs. General Greene, at whose house he died (four years after her own death), and by whose side he lies buried, in a little coquina-walled graveyard hidden in the depths of an olive grove and surrounded by tropical fruits and flowers.* Somerville is the county-seat of Somerset, and is the trading center of a large agricultural district. When a Somerville citizen wishes to especially emphasize his unwillingness to perform some act, he asseverates that he "wouldn't do it for a farm." This is a great horse-breeding country, being well watered, and clothed with a mantle of soft, succulent grass, a pretty feature of the landscape being the Raritan River.

---

*See "The Story of an Old Farm," by A. D. Mellick, Jr.

BUTTERMILK FALLS.

The conversation around the stoves in the various hostelries turns upon horses. The men either have been to see a horse or are going to see one the next day; and a quaint old farmer remarked, as a long funeral passed the hotel, that, if the mourners had turned out for the deceased's gelding "who'd well nigh gotten down to 2.18 just before he broke down and had to be shot," instead of for the deceased himself, he "could ha' understood it better." If the talk drifts back to old staging days, as it is sure to do, you hear of the driver, living still in hale and hearty old age, who would beat the rival stage, no matter how many horses he'd kill; who once drove sixteen horses from Elizabeth to Newark on a wager, and who of a winter night would harness up eight horses to a sleigh and take the young folks for a spin over the crisp, moonlit snow.

Early in October the Somerset County Fair attracts people from all over the surrounding country to Somerville, and horse-talk is somewhat varied with pumpkin and butter lore.

There are excellent stores in Somerville, first-rate schools, and many pretty dwellings; and, with its charming location, among rolling meadowland watered by rills from the hills to the north and by the Raritan, it forms a fitting terminus to the chain of delightful cities, towns and villages, which the suburban system of the Central Railroad of New Jersey links together.

From Somerville a branch railroad runs through a rich agricultural district to Flemington, which is to Hunterdon County what Somerville is to Somerset. The intermediate stations are Roycefield, Flagtown, Neshanic, Woodfern, and Three Bridges.

While the American army was in camp at Bound Brook, Washington had his headquarters at a house in Somerville, now known as the Caleb Miller house. It is a plain white clap-boarded dwelling, with a porch supported by two old-fashioned fluted columns. It has a large square hall overlooked by a long gallery formed by the landing of the staircase. The only Revolutionary relic is in the room which Washington and his wife occupied. This is a fire-place framed with tiles on which Biblical scenes are depicted in blue and white.

RARITAN.—A short distance above here the North and South Branches of the Raritan unite, and the river is tapped by a water-power company, which supplies some half a dozen manufactories. These are flourishing and give some indication of the future prosperity of the place. The brick house on the north bank of the Raritan was built long before the Revolution, but, while most charmingly old fashioned outside, the interior has been completely stripped of everything old. A debtor, who was hounded by creditors, once occupied this house. He had painted on the outside of the doors of four of the upstairs rooms respectively "New York," "Boston," "Philadelphia" and "Washington," and, when a creditor called, he would be informed that the man he wanted to see was in "New York," or another of the places mentioned according to the room in which he had secreted himself.

SIMCOE'S RAID.—Simcoe's raid, one of the most dashing exploits of the Revolution on the British side, extended as far as Raritan. It was daringly conceived and brilliantly executed almost to the end, when it was overtaken by disaster. Simcoe's force consisted chiefly of American Tories, known as the "Queen's Rangers." October 26, 1779, the party landed at Elizabeth. The object was to proceed swiftly to Van Veghten's bridge, over the Raritan near Finderne, there to destroy the flat-boats left by Washington; to make a circuit around New Brunswick, then show themselves to the Americans, and fall back toward South River Bridge where troops under Major Armstrong lay in ambush, ready to capture

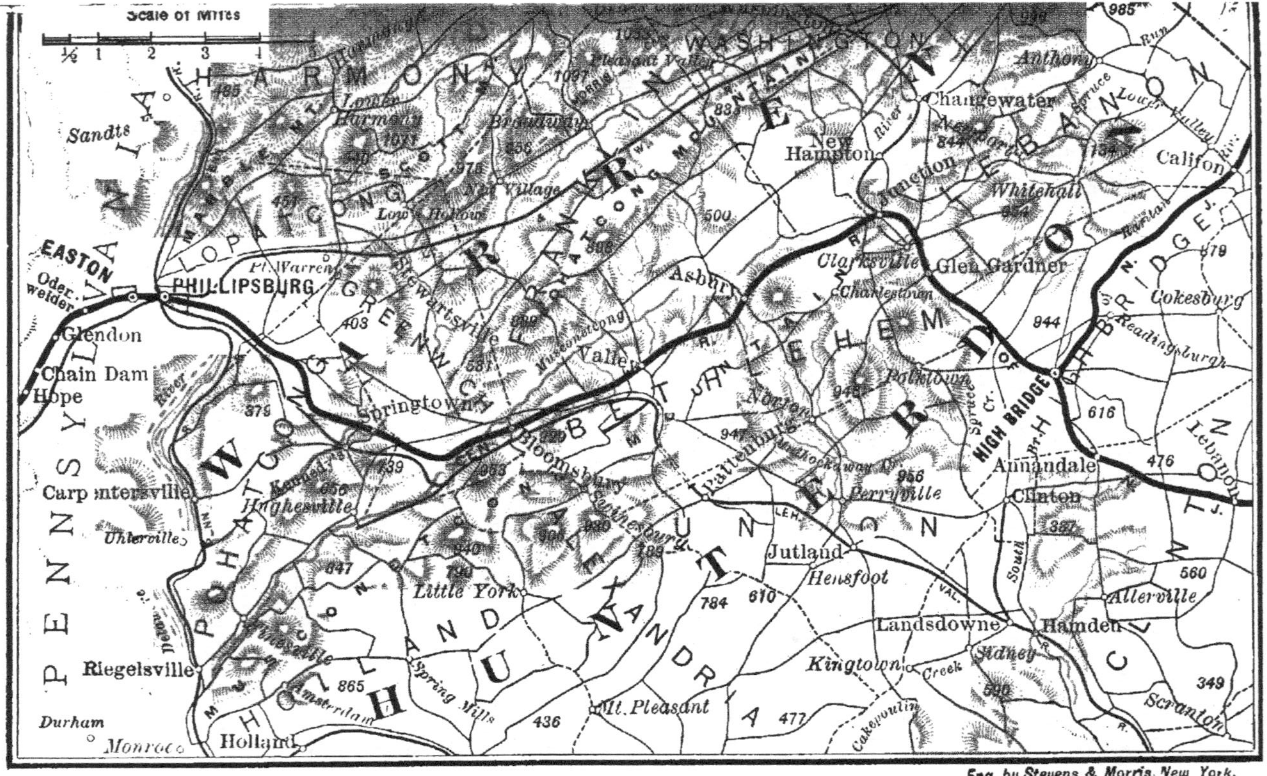

Eng. by Stevens & Morris New York.

the pursuing Americans. In order to avoid having the Americans at New Brunswick receive notice of the raid Simcoe's men, being accoutred similarly to Lee's Legion, pretended that they belonged to the American army and were in pursuit of a party of Tories. They were so daring as to stop at a Continental forage depot, assume the character of Lee's cavalry, wake up the commissary about midnight, draw the customary allowance for forage, and give the usual vouchers.

A country lad, deceived by this ruse, led the party to one of Washington's old camps at Bound Brook, but Simcoe found it impracticable to burn the huts. At Van Veghten's bridge they set fire to and destroyed eighteen boats, burned the old Dutch meeting-house built in 1721 at Raritan, and the Somerset County Court House. But in manœuvering to make a circle around New Brunswick, they missed the particular cross-road Simcoe had planned to take. The country having meanwhile become alarmed, small detachments of militia were hovering around him. One of these detachments formed an ambuscade near DeMott's tavern, two miles west of New Brunswick, and firing upon the Rangers, killed Simcoe's horse, made a prisoner of Simcoe himself and dispersed his followers.

RARITAN TO PHILLIPSBURG.—From Raritan to Phillipsburg the settlements have few interests other than local. The country, however, becomes more mountainous and the scenery more romantic and picturesque. The stations are at North Branch, White House, Lebanon, Annandale, Clinton, High Bridge, Glen Gardner, Junction, Asbury, Valley, Bloomsbury and Springtown.

At White House the railroad commands a view of Cushetunk mountain, south-west of the tracks, and runs along its northeastern base to Lebanon. This is a curious ridge, shaped like an oval dish with a deep depression between the sides, completely closed except to the west, where, however, two hills nearly complete the circuit. This mountain incloses a nest of pretty farms in what is aptly named Round Valley.

The whole outward slope of Cushetunk offers fine opportunities for country residence, especially for gentlemen stock-farmers, as they can have their pastures on the rolling meadows, and their residences on the mountain side.

From White House the Rockaway Valley Railroad connects with the Jersey Central, opening up a section rich in lime. About half a mile west of the station and a little north of the railroad is a stone house, bearing date 1757, and the initials C. V. H. This was built by a German "redemptioner," one of the "white slaves" of New Jersey, who bound themselves for a certain number of years to whoever paid their passage money. The vicinity of Annandale (Clinton) is also rich in lime.

At High Bridge the North Branch of the Raritan descends from German Valley through a gorge, which, for wild beauty, has not its equal within the same distance of New York, nor perhaps for many miles further around the city. Excepting Lake Hopatcong, it is probably the most romantic feature of the inland landscape of New Jersey. It is so strikingly beautiful that the railroad company at one time intended to lay out excursion grounds through the glen, but finally decided upon Lake Hopatcong as more desirable. The High Bridge branch of the Central Railroad of New Jersey ascends to the level of German Valley through this gorge.

The river furnishes fine water-power at High Bridge. Since 1758 it has been utilized in the Taylor Iron Works. Here cannon balls were cast for the American army during the Revolution, and carted to Trenton, New Brunswick and Philadelphia. Part of the Taylor mansion was built in 1725. One of the rooms was occupied, during the Revolution, by John Penn, the last Colonial Governor of Pennsylvania, and his Attorney-General,

Chew. The manufacture of plumbago is extensively carried on here.

Between High Bridge and Glen Gardner is another stretch of picturesque country, the railroad looking down on its southerly side upon a glen through which the little stream of Spruce Run speeds on its way towards the North Branch of the Raritan. At Junction connection is made with the Delaware, Lackawanna and Western Railroad. A short distance above Junction the railroad swings into the lovely Musconetcong Valley and passes through Asbury, Valley, Bloomsbury and Springtown to Phillipsburg. About a mile below Bloomsbury, in the bed of the Musconetcong, is Butler's Hole, said to be sixty feet deep. In a huge boulder opposite are several mould-like depressions. There is a tradition that here Spanish buccaneers melted their ill-gotten spoils into ingots, which they hid in the bottom of the river.

Phillipsburg was named after the Indian King Philip. It is beautifully situated on the east shore of the Delaware. Here was a favorite fishing-ground of the Indians. Mount Lebanon and Reese's Rock (Mount Parnassus), command superb views of the river.

For a long time Phillipsburg was a struggling suburb of Easton, but, since it has become a focal point for several important railroads, which penetrate into the coal and iron fields of Pennsylvania, it has become a flourishing place of manufacture, especially in iron.

From here the Central Railroad of New Jersey crosses to Easton, entering there on its Pennsylvania division, which leads to Bethlehem, Allentown, Siegfried, Mauch Chunk, Wilkesbarre, Scranton and Tamaqua—the rich coal and iron fields of Pennsylvania.

CHAPTER V.

SCHOOLEY'S MOUNTAIN.

From High Bridge the Central Railroad of New Jersey sends a spur to the north into the very heart of the mining section of New Jersey. Ascending the beautiful gorge of the North Branch of the Raritan River, it enters German Valley, which lies between Schooley's Mountain and the Chester Hills. This is a rich agricultural district and especially devoted to peach-growing. From the first station, Califon, 150,000 baskets of peaches are shipped every season. On the mountain side much hard wood for wagons and interiors is cut, portable saw mills being used; and a wall on one of the finest places on Rumson Neck is built of granite from one of the several rich quarries reached by this branch. There are also extensive lime quarries in the valley, especially at Vernoy, the second stop. At Middle Valley are several large creameries. From German Valley a branch runs to Chester and Hacklebarney, the former the center of a large peach-growing district, the latter a mining settlement, the mine machinery being worked by water power.

CHESTER.—The village of Chester was formerly called Black River, after a tributary of the North Branch of the Raritan. Under the reign of Queen Anne, about 1713, this tract was settled by emigrants from Long Island, among whom were the ancestors of Lincoln's Secretary of State, Wm. H. Seward. Tradition confers special distinction upon one Deacon Fairclo as the father of twenty-one children;—the climate is still as invigorating as it was in those days.

At the time of the first settlement there were no turn-

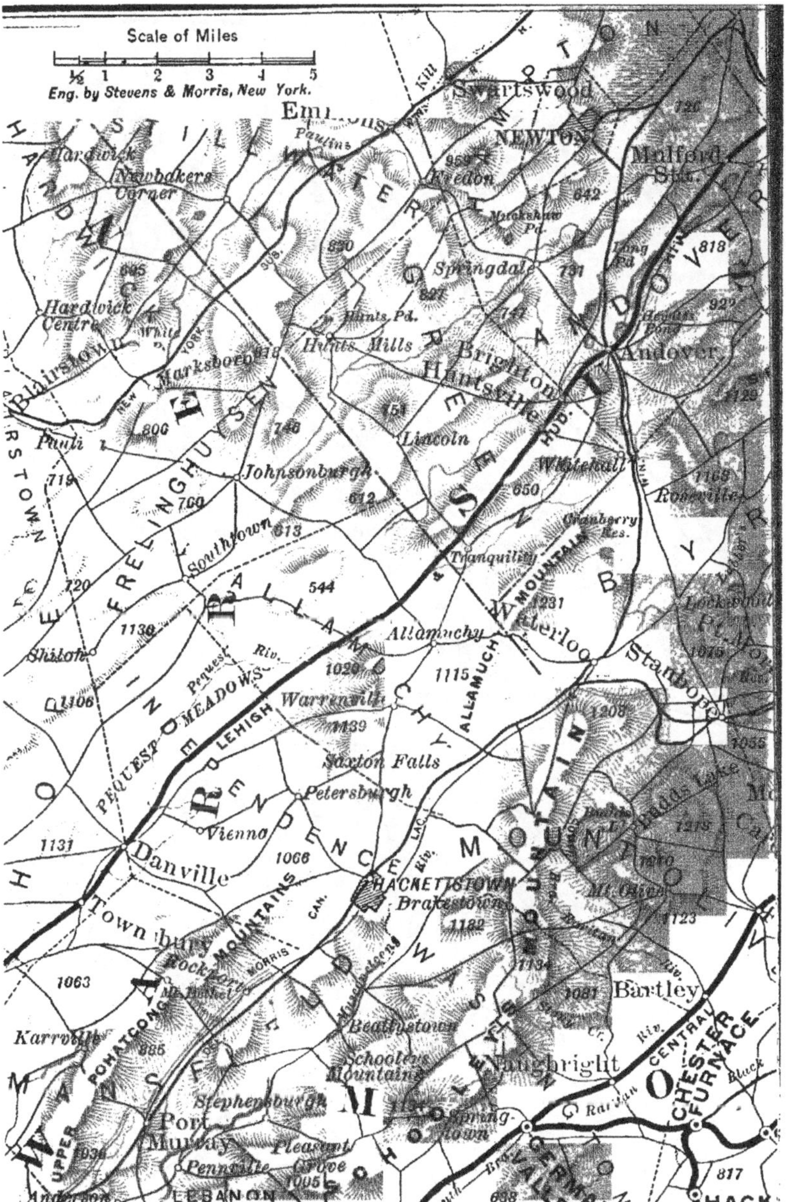
Scale of Miles
½ 1 2 3 4 5
Eng. by Stevens & Morris, New York.
Swartswood
NEWTON
Hardwick
Newbakers Corner
Hardwick Centre
Blairstown
Marksboro
Fredon
Springdale
Hunts Mills
Huntsville
Brighton
Andover
Lincoln
Johnsonburgh
Whitehall
Roseville
Southtown
Tranquility
Cranberry Res.
Allamuchy
Waterloo
Stanhope
Shiloh
Warrenville
Saxton Falls
Petersburgh
Vienna
Danville
HACKETTSTOWN
Brakestown
Mt. Olive
Townsbury
Rockport
Bartley
Karrville
Beattystown
Schooleys Mountain
Naughright
Stephensburgh
Port Murray
Pleasant Grove
Pennville
Anderson
LEBANON
THE POINT
CHESTER FURNACE
HACK
GERMAN VALLEY
PEQUEST MEADOWS
LEHIGH
ALLAMUCHY MOUNTAIN
POHATCONG
MOUNTAINS
MORRIS CAN.
UPPER
CENTRAL

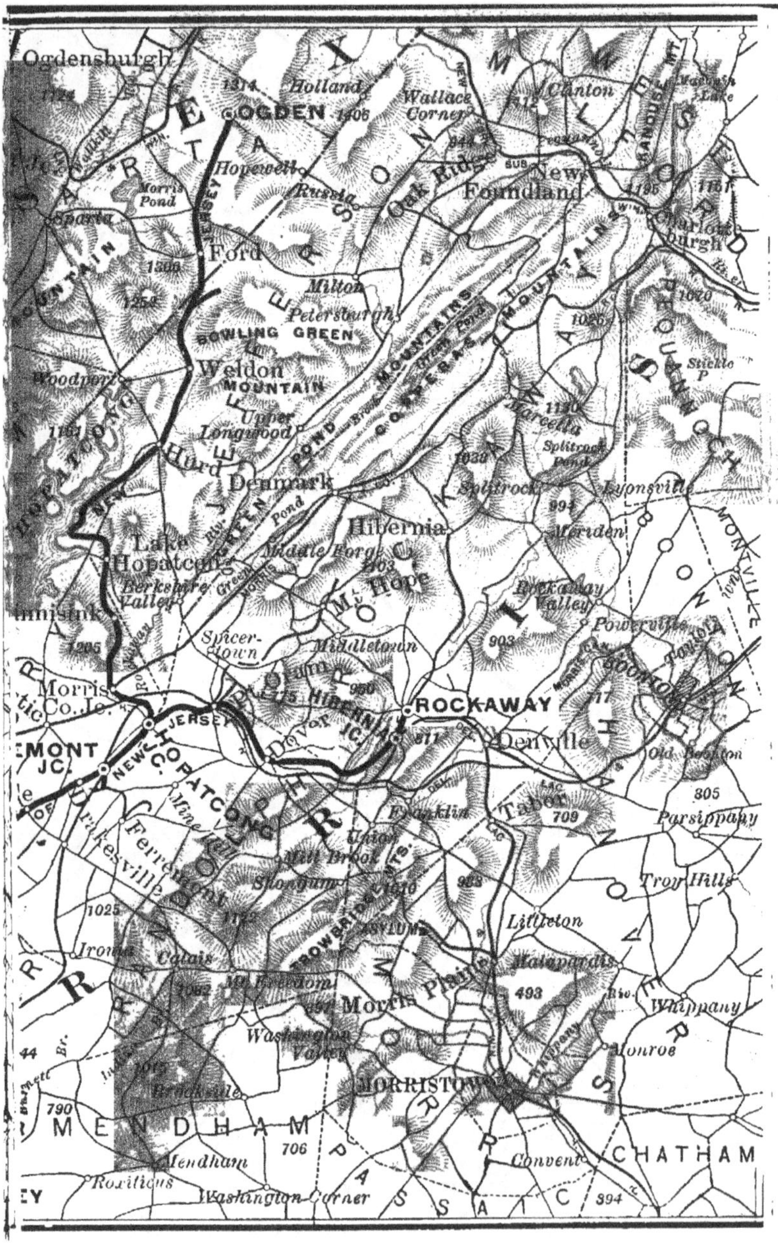
Ogdensburgh
OGDEN
Holland
Wallace Corner
Clinton
Hopewell
New Foundland
Charlotteburgh
Morris Pond
Sparta
Ford
Milton
Petersburgh
BOWLING GREEN
MOUNTAIN
Weldon
Upper Longwood
Hurd
Denmark
Hibernia
Lake Hopatcong
Berkshire Valley
Middle Forge
Mt. Hope
Marcella
Splitrock Pond
Splitrock
Lyonsville
Meriden
Rockaway Valley
Powerville
Spicertown
Middletown
Morris Co. Jc.
ROCKAWAY
HIBERNIA JC.
Dover
Denville
Old Boonton
Parsippany
Franklin
Tabor
Ferromont
Mill Brook
Shongum
Troy Hills
Littleton
Ironia
Calais
Mt. Freedom
Morris Plains
Malapardis
Whippany
Washington Valley
Monroe
MORRISTOWN
MENDHAM
Mendham
Brookside
Convent
CHATHAM
Washington Corner
PASSAIC
MORRIS

pikes, the travel being by bridle paths over the hills through dense woods. As late as 1768, Rev. William Woodhull, who opened a school which was attended, among others, by Mahlon Dickerson, who became Secretary of the Navy under Jackson, entered Black River on horse-back with his wife and child riding behind him. As early as 1745 and 1747 respectively, a Presbyterian and a Congregational Church were erected, the first pastor of the former being Rev. Samuel Harcour, who, for doctrinal errors, was deposed from the ministry in 1763, the case being cited by Hodge as an interesting illustration of early Presbyterian rigor.

The hills about Chester are rich in deposits of magnetic iron ore which, being low in phosphorus, is valuable for the manufacture of Bessemer steel. The forge at Hacklebarney has been in operation for more than a century.

From German Valley, Schooley's Mountain (p. 76) is reached by stage. German Valley was settled by exiles from Saxony. When Frederic August, Elector of Saxony, renounced Protestantism in 1697, many of his subjects removed to Neuwied, Prussia; thence to Holland; and in 1707 to America. The vessel's course was laid for New York, but adverse winds drove it into Delaware Bay. The emigrants reached Philadelphia and started for New York, but impressed with the fertility of German Valley they remained there. The Lutheran Church has existed since 1745. In the old church there was originally no fire-place nor chimney, just a hole in the roof, the fire being made on the floor, and in murky weather the congregation had difficulty in determining whether it was suffering more from the long sermon, or from the smoke which filled the building.

In the North Branch of the Raritan, near Naughright, there is fine trout fishing. At Bartley's are a well-known manufacturer of turbine wheels and a saw mill; and also good creameries. Fire sand is found in abundance at Flanders and Cary's. This is used in puddlingfurnaces because, as it is infusible, it catches the scrap. From Drakesville the railroad traverses the plains of Succasunna to Kenvil, where there is an important dynamite factory, which does business as far west as the Rockies; another establishment of the company supplying the Pacific Coast.

On the mountains, on either side of the German Valley, there is excellent quail and partridge shooting. At Flanders, which is the station for Budd's Lake (p. 86), and a good trout stream comes down Schooley's Mountain.

At Kenvil there is a junction for Lake Hopatcong and the Ogden Mine Railroad; and for Port Oram, Dover and Rockaway, where connection is made with the Hibernia Mine Railroad; while, between Kenvil and Lake Hopatcong, connection is made with the Morris County and Green Pond Mine Railroad for Lake Denmark, Green Pond, and the Mount Hope Mines. Denmark is a pretty sheet of water among the mountains, and Green Pond boasts one of the noblest features among the Highlands. The mountain seems to have been cleft in two, one side thrown over to the east, while to the west there remained standing a sheer, high wall of trap rock, three miles long, and at points almost as grand as the Palisades of the Hudson.

Port Oram is a recent settlement, dating back only to 1868. It is actively engaged in mining and manufacturing. The famous Richard Mine yields 65 per cent. of metallic iron, which runs in a vein sixteen feet broad. There is also a silk-mill where silk mufflers are made.

Dover is the center of the mining district. It also manufactures mining machinery, which is shipped all the way to Colorado. Silk-mills are also in operation there. It is on the Rockaway River, which supplies fine water-power. On the site of Dover there were already in 1722 a forge and several dwellings. A Quaker meeting-house stood there in 1748 or even earlier.

At Rockaway there are a foundry, a factory for rock-crushers, stampers and separators, and also knitting-mills. The places named, besides being flourishing manufacturing localities, are pleasantly situated for residence.

The iron industry of Morris County dates back into the seventeenth century, at least by implication. In 1714, the tract embracing the Dickerson Mine was taken up because of its minerals from the proprietors of West Jersey by John Reading; and tradition affirms that previous to that date, probably as far back as the latter part of the preceding century, there were forges in operation, whose owners helped themselves to the ore without stint or charge. Indian arrow-heads and utensils made of iron, show that the aborigines understood its uses, and they had given the name Succasunna, meaning "black" or "heavy stone," to a district near the mine where the village of Succasunna still flourishes.

The first forge of which we have somewhat exact knowledge was built in 1710, at Whippany, whither the ore was brought from the Succasunna district in leather bags on horse-back.

On what is still called Jackson's Brook, in Dover, a forge was built by John Jackson in 1722, other forges having meanwhile started up at Morristown. This Dover forge, or the "Quaker Iron Works," as it is called in a deed of 1743, became in 1761 the property of Josiah Beman. About 1730 a forge was built at Rockaway by Jacob Ford, who subsequently built also at Mt. Pleasant and Denmark, and at what became known as Middle Forge, half-way between these two. The Rockaway forge still exists.

Weldon Forge was built about 1800 by Moses Hopping; Hurd Forge by Daniel and Joseph Hurd about 1795. There were also forges on the North Branch of the Raritan, near Budd's Lake, Bartley's and Flanders, but they have long since gone to decay.

The Dover Iron Company is the third oldest slitting or rolling-mill erected in Morris County. Between 1749 and the breaking out of the Revolution, there was a prohibitory act of Parliament. Nevertheless, about 1770, David Ogden began carrying on slitting in secret at Old Boonton, the works being underneath a grist mill. Gov. William Franklin visited the place for the purpose of investigating the alleged fraud, but as Ogden took the precaution to dine the Governor before he went over the premises, the representative of His Majesty actually waxed indignant at the unfounded slander. Next in point of time was the famous Speedwell Mill near Morristown, where subsequently Professor Morse and the Vails made so many interesting experiments in putting the former's telegraphic theories into practical operation. The mill at Dover was erected in 1792 by a firm which purchased the Beman Forge. The Rockaway rolling-mill was built in 1822.

SCHOOLEY'S MOUNTAIN.—Schooley's Mountain is a range some sixteen miles long, 1,200 feet above tidewater, and overlooking the Musconetcong Valley on the north and German Valley on the south. It has the peculiar characteristic that, instead of rising in peaks, its top is a plateau of rich farm lands and forest, averaging one and a quarter miles in width. Beautiful views may be had from various points on the edge of this plateau, and also from the road crossing the mountain between German Valley and Hackettstown, and running through the village of Schooley's Mountain and past the noted Chalybeate Spring. This road is a branch of an old post-route between New York and Easton via Elizabethport, and this point for crossing the mountain was doubtless selected because of a slight depression in the plateau. The draught of air through this depression and the altitude combine to make Schooley's Mountain a pleasantly cool resort, and, as the air is dry and bracing, and the Chalybeate Spring very effective in certain diseases and invigorating in all cases, the place is a health resort as well as a summer retreat.

Schooley's Mountain Village is one of the oldest summer resorts in the United States. Indeed, it was a health resort before there was a United States, for the Chalybeate Spring was famed already among the Indians for its valuable curative properties. The Pennsylvania tribes sent for its waters, and Tedyeesung, the renowned king of the Lenni Lenapé, is said to have always kept his camp-fires burning within three miles of it, in order that he might resort to it at any time. There is also a tradition that the spring became known to the whites only through chance, the Indians keeping its existence a profound secret, and the whites first learning of it through a hunter, who, coming upon it, quenched his thirst from the rill and, noticing the peculiar mineral taste, reported his discovery.

It is certain that there was a hotel here as early as 1795. The old building still forms part of the Heath House. It is appropriately called the "Alpha." In it are still several old mirrors and pieces of furniture, relics of the hotel accomodations of an American summer resort of the last century.

When the road across Schooley's became a regular turnpike and post-route in 1809, the mountain was one of the most famous summer resorts in the United States. It is spoken of by the French scientist, Milbert, in his "Itineraire Pittoresque du Fleuve Hudson et des Parties Latérales," the author having made his trip in 1815. The description is embellished with two engravings, one of the rock from which the spring flowed (there was then no spring-house or basin), the other of the cataract, still a natural feature of great beauty. In a circular issued by the Heath House in 1828, the proprietor offers as one of the chief inducements to visitors the "opportunity of associating with company the most gay and fashionable"; and this statement is borne out by the old registers preserved in the hotel office in which the names of the leading New York, Philadelphia and New Jersey families of the day are found. These traveled in their own carriages with their servants and domestic pets, usually taking two days for the journey, though the trip from New York to Easton could be made in one day by stage, as appears from an advertisement of McCowy, Drake & Co., April 26, 1828, who advertised their stages "to run through in one day and by daylight" from New York to Easton via Elizabethport, Morristown and Schooley's Mountain Springs. People of moderate as well as those of large means came to the spring, the former pitching tents or erecting temporary shanties in its vicinity, so anxious were they to take advantage of its curative properties.

The spring is about half a mile from the hotels and Schooley's Mountain Village. It is on a high rock to the right of the road from Hackettstown, and the water now led through a pipe into a basin around which a summer-house has been built. Glasses can be obtained for a small fee in a neighboring house, but visitors are advised to take them from their hotel or cottage. The water can also be ordered at the hotels. It is especially recommended for calculus, kidney complaints, torpor of the liver and as a tonic.

It was analysed early in the century by distinguished chemists, who also testified that it was the purest chalybeate water in the United States. Following is the latest analysis by Dr. T. M. Coan :

Solids.	*Grains per gallon.*
Sodium bicarb.	0.58
Magnesium carb.	1.60
Iron carb	0.58
Calcium carb.	1.42
Culcium sulph	1.68
Alumina.	0.14
Silicic acid	0.74
Sodium chloride.	0.43
	7.17

with a trace of Manganese Carbonate and of Ammonia.

The "Alpha," mentioned above, was probably a road-house, flourishing on the patronage bestowed upon it by the passengers of the stages which followed the post-route across the mountain. It is even said to have been a "jug-tavern," similar to those which in olden times flourished among the Jersey Pines, and which owed their peculiar name to the fact that their whole stock in trade consisted of a jug of apple-jack, from which, however, any liquor called for by a customer was poured. It was simply apple-jack under another name. It was thus possible for a wood-ranger on Schooley's Mountain to be all his life long the victim of a spirituous delusion—a condition of affairs which, however, does not now exist even in the most remote corner of New Jersey.

As the Schooley's Mountain Spring became more famous, the jug-tavern and road-house improved in character. Additions were made until some 350 guests could be accommodated, and the original little building seemed so remote an object in history, that it was dubbed the "Alpha," pretty much as if it were the very beginning of creation. It is thought to be, and probably is, the oldest summer-resort building in the United States.

The buildings of the Heath House are all old-fashioned and ample, standing in spacious, pleasantly shaded grounds—some twenty-five acres in extent. Near the main entrance to these grounds, to the right walking from the house, is a group of huge boulders, all of striking shapes and one of them appropriately named the Devil's Arm-chair. Tradition says that the Indians often gathered here in council, the Chief presiding in the Devil's Arm-chair. The Heath House is comfortably furnished; the table is plain but plentiful. It makes no pretence of affording fashionable amusement, but seeks rather to attract those who find recreation in restful quiet, and are satisfied with home-like accommodations. Among its guests are several who have made it their summer retreat for over thirty years. Perhaps the fairest idea of the character of the accommodations, service and table can be given by saying that these are excellent for the prices charged, which are as follows: $12.00 to $14.00 each adult per week for single rooms; double rooms, two persons, $24.00 to $28.00 per week ; double rooms, one person, $18.00 to $21.00 per week; one week and less than two weeks, $2.00 per day; $7.00 per week for children and nurses taking their meals at the children's table; for children under twelve years of age, taking their meals at the public table, $10.00 per week; all over twelve years, full price. Nurse with infant, occupying room on guest's floor, will be charged as one person. Transient guests, $2.50 per day. Livery, bowling-alleys and billiard-room are connected with the hotel, and there is a tennis-court. There is a stage for private theatricals, and a dark-room for amateur photographers. Rates of carriage hire are as follows: Budd's Lake, $6.00; Hackettstown, $3.00; double team, $1.50 an hour; single team, $1.00 for first hour, 75 cents an hour afterwards. Stage between hotel and either railroad station, 50 cents.

Another hotel, the old Belmont Hall, changed hands last spring, and is now called the Dorincourt. It is a fine building, and its proprietors purpose to cater more to the fashionable element. It was not completed until late last season, and could not be put in good running order; so it would hardly be fair to rate it in this edition.

Schooley's Mountain has some peculiar characteristics. The hotels are 1,200 feet above the sea, the air is dry and cool, especially at night, yet the immediate surroundings do not offer the slightest suggestion of a mountainous district. This is due to the fact that the land does not rise to a peak but to a broad plateau, given over to farms and woodland, so that it has the appearance of a fertile agricultural district rather than of a mountain. It is only from points along the edge of the plateau which command views over the Musconetcong or German Valleys that one can realize the elevation. The hotels, having begun as road-houses, were built at points along the post-route over the mountain where it would be convenient for the stages to stop. The idea of locating them where they would command one or another of the beautiful vistas to be

CATARACT—SCHOOLEY'S MOUNTAIN.

had almost within a stone's throw, did not occur to their proprietors. They were built about half-way between German Valley and Hackettstown, where the road is fairly level and the stages could easily stop, enough time having elapsed since they left the valley for the driver (certainly) and some, if not all the passengers, to have become thirsty and possibly hungry. Hence there are many points on the edge of the plateau which, though they command superb views and seem to have been especially designed for summer hotel sites, are still unoccupied. Until these have been visited, a sojourner at Schooley's Mountain does not begin to realize its attractions; but, as he gradually discovers them and the gushing mountain streams and water-falls within easy driving and walking distance from the hotels, he begins to appreciate the fact that no resort so near New York offers such a variety of mountain scenery. Moreover, as several of these spots are known to but few, whoever will start out in search of Basin Rock, the Point or "Pint" Mill, Eagle's Nest, Bald Mountain, Prospect Hill, the Cataract, or Striker's Falls, can do so with something of the importance and zest of a discoverer. Sitting on a hotel piazza reading a pink, blue or yellow-covered novel; riding two or three times around "the circle"; taking the least interesting of the drives about Schooley's Mountain—that to Budd's Lake—because it seems to be the only one anybody knows anything about; strolling down to the spring-house and back—doing these and similar things is not the sum of enjoyment one can derive from a visit to this resort. Yet, but little more is done—and no wonder; for the very people who are most interested in making the beauties of Schooley's Mountain known to the hotel guests, do not themselves know even of the existence of these attractions.

A lovely glimpse of the valley of the Musconetcong is

had from a point called Valley View, barely more than 200 rods in front of the hotel. Here you stand at the apex of a clove running up from the valley. It is a narrow opening, intercepted here and there by wooded promontories, and musical with the rushing waters of a brook completely hidden from view by the foliage. At the foot of the clove lies the Musconetcong Valley, with the steeples of Hackettstown peeping out from among the trees that shade its streets, the hills beyond forming a picturesque background with their fertile slopes, here yellow with grain, there green with corn, and dotted with white farm-houses or red barns. The point from which this view is obtained is shaded by a clump of trees, growing up among boulders whose gray tones, lit up now and then by glints of sunshine, harmonize with the cool shadow of the foliage; and, to one looking out from this recess, the glimpse of valley and distant hills, at the end of the soft, green slopes of the clove, seems unusually bright and friendly.

Another point of interest of easy access from the hotels—about a mile and a half—is the Cataract. To reach it take the Hackettstown road to a point a little below the Spring House, where a rough wood-road enters the woods to the right and crosses the brook. Follow this road, always keeping to the right, until a second brook is reached. A little way up this is the Cataract. Here there is an almost sudden descent of about 100 feet from the plateau into the rocky clove up which we have followed the brook whose waters now come leaping down from boulder to boulder, sending their white spray flying in showers, rushing through crevasses, frothing up against the trunks of fallen trees and finally hurrying away through the clove toward the valley of of the Musconetcong. The ascent of the Cataract is best made on the left. Near the top is a large flat rock. On stepping out upon this one obtains without the slightest previous intimation a glimpse of the valley similar to that had from Valley View, the rock overlooking the tops of the trees at the foot of the Cataract. The view seems the lovelier for being had so unexpectedly from the very heart of the forest. Not far from the top of the Cataract are fields through which one can easily reach the village and hotels, so that it is not necessary to again descend into the clove and return by the Hackettstown road.

There is another and even more picturesque waterfall near Schooley's Mountain. This is Stryker's Falls, off the German Valley road, and is reached by following this road through Springtown to a stone-quarry, and there turning off to the left a short distance into the woods, from where one is guided to the falls by the sound of rushing water. As the path is not, however, easy to find, it is well to get a boy in Springtown to act as guide.

About one and a quarter miles from the hotels is Prospect Hill. The road leads down into a ravine and then up a steep hill. In a pasture-field to the right is a chestnut tree, from near which one obtains a view of the Musconetcong Valley, less circumscribed than that from Valley View, and enhanced by the delicate hues of the Blue Ridge in the distance.

A fine view of the Musconetcong Valley and beyond to the Delaware Water Gap, the gap in the mountains being clearly defined, is to be had from Mr. Alfred Sully's place, on the road between Drakestown and Hackettstown. Almost the same view can be had also by taking the Budd's Lake road almost to Drakestown, but turning from it on to the road which, near Drakestown, goes off to the left. Very soon after getting on this road the view referred to is obtained. The road eventually leads into that crossing the mountain, which may be taken back to the hotels, but, as the map shows, there is a way of avoiding the steep grade of the mountain road, by taking the first turn to the left, which brings one on to the Budd's Lake road. Fine views can also be had from Eagle's Nest, Bald Mountain and Drake Hill ; but they do not differ in character from those already described. Each of the points named, excepting Valley View, which is too near, and the points near Drakestown, which are too far (for the ordinary walker), is a pleasant excursion for a morning or afternoon. It is also an easy matter to drive to Budd's Lake and back in half a day, but it is better to take a day for this purpose, in order to enjoy the boating and fishing on this attractive sheet of water.

A superb panoramic view is had from a huge rock at the edge of the plateau, about seven miles southwest of the hotel by the shorter road. This grim reminder of a remote geological age of ice and gloom is variously called Eagle Rock, Basin Rock, and the Point, the last name being applied to it on the map of the Geological Survey of New Jersey. The view is undoubtedly the finest to be had from any part of Schooley's Mountain, and the roads to it (for there are two) also afford many glimpses of pretty scenery, and at least one exceptionally beautiful vista. Of course it is an easy morning or afternoon drive; but it is a delightful day's excursion afoot, and, if the party wants to walk only one way, the train can be taken at Port Murray, only one and a half miles from the Point, for Hackettstown, and the stage from there for Schooley's Mountain. The Point is reached by the Pleasant Grove road. Just before it descends towards "the Grove" one has to the right a view extending, on a clear day, to the Water Gap. After passing through Pleasant Grove, two courses are open—to take the first road to the left and swing around past Mount Lebanon Church to the

Point, or to proceed on through Pennville toward Anderson, near which latter place a r ad to the left leads to the Point. A glance at the map will show that one can take either of these routes, and by simply continuing on past the Point return by the other, the road forming a loop to the Pleasant Grove road.

Both routes are about equally attractive, but parties afoot who intend taking the train from Port Murray will, perhaps, find greater variety along the Mount Lebanon Church road.

The Point is not reached by the wagon road. It is necessary to clamber for about ten minutes up a steep path so overgrown with brush that it is advisable to have a boy from one of the families on the mountain-side to act as guide. The view from the rock up and down the Musconetcong Valley is superb. Along the foot of the mountain flows the Musconetcong, whose course is marked by the sinuous line of trees which shade its cool current. Now and then, through a break in the foliage, its glistening waters come into view only to vanish again under the green archway. Rich pasture lands and fields of waving green impart a velvety softness to the slope of the opposite hills, whose predominating colors are varied with the white and red of neat farm-houses and ample barns. The series of mountain ranges beyond fade away from dark green to delicate tints of blue which finally lose themselves in the hazy distance. As with all the views in this region, there is nothing rugged or grand in this. But it has a certain feminine softness and grace which give it a peculiar charm. On a bright day Nature is seen here in one of her most affable moods; and, even if a storm be brewing, she is, perhaps, all the prettier for her passing petulancy.

At the foot of the mountain is the Point Mill, and from the bridge the view up the Musconetcong is one of tranquil beauty—the water as it flows over the low dam stretching like a band across the stream, with soft borders of grass along the shaded banks above and floating islands of lily-pads beyond.

BUDD'S LAKE nestles delightfully among the hills. It is a pretty sheet of water, full of bass and pickerel. In season there is abundant shooting over the mountain ; and in the early fall excellent duck-shooting on the lake. It is reached from Flanders.

Joseph Bonaparte once thought of settling here, but during the negotiations he chanced to discover a caricature of his illustrious relative, belonging to the proprietor's daughter, which caused him to change his mind.

CHAPTER VI.

LAKE HOPATCONG.

Etched in silver, in the wooded slopes of the Highlands of New Jersey, lies Lake Hopatcong. From among the hills it greets the beholder with one of nature's friendliest smiles. Its aspect is nearly always cheerful, for its surface is rarely agitated beyond a ripple. Yet this placid and ingenuous looking sheet of water has come down to us from one of the grimmest epochs in the earth's history. Once there doubtless flowed through the valley which now encloses Lake Hopatcong a little stream, a mere thread of silver, winding through a strip of meadowland. Then, the giant forces of the remote North having gathered themselves for a slow but fearfully sure advance, a glacial army crept southward for untold thousands of years, crushing and levelling all obstacles and pushing in its van a veritable mountain range of debris—which was to its vast mass no more than the scum of the strand is to the sea. On and on it came, until, in what is now the State of New Jersey, it met an opposing force greater than itself. Evidence of the destruction that ensued is scattered all over that portion of the State which the ice had covered; for, as the glacier melted away to the north it left in its retreating track the debris it had gathered in its advance. Right at the outlet of the valley through which our little stream had flowed, the glacier, on its retreat, heaped up a dam of drift; and, when the ice had crept back so far northward that the hills again wore a friendly summer aspect, they held in their embrace one of the loveliest lakes in the world. How long it mirrored their soft contour before it gladdened the eye of man no one can say. But it is certain that it was known to the Indians long before the whites set foot upon this continent. Its Indian name, Indian relics found along its shores, and the fact that it lay in the track of a well-known Indian trail, substantiate this.

Hopatcong (originally Hopatchung) signifies "Pipe Water," a name descriptive of the shape of the lake before its waters were artificially raised for the purposes of the Morris Canal in 1832. Since that time it has made islands of several promontories and entirely covered others, flowing far back into the recesses of the hills, and flooding the marsh which once divided it from Little Pond, with which it now forms a lake some seven miles long, and two miles in width at its broadest point. The main body of the lake is barely a mile wide at the most, but the coves on its western shore (the River Styx, Byram and Henderson Coves) are deeply indented in the hills.

The Indians who dwelt on the shores of Lake Hopatcong were the Nariticongs, a branch of the Lenni Lenapé. Fish were plentiful in the lake and game abounded among the surrounding hills. The site of their principal settlement is now submerged. It was near Halsey Island, which, before the damming of the lake, was connected with the mainland. The site was easily located by circular hearths of fire-blackened stones, from whose number it was judged that the Indian village was formed of some fifty wigwams. The dead were buried on the extreme end of the promontory, now Halsey Island, where many of their remains have been dug up. Historians have also mentioned an Indian causeway of stone connecting Bertrand Island with the mainland, but no trace of it can now be found. It is also stated that workmen, while digging the Morris Canal, exhumed, near the

outlet of the lake, portions of an Indian skeleton with an armbone measuring thirty-six inches, which, if it was in proportion to the rest of the body, would indicate that this ranger of the primeval forest rejoiced in an immense frame. For he must have been at least eight feet in height. If a Nariticong, the race was indeed a noble one. Who knows, however, but that this skeleton may have been a relic of prehistoric man, borne down from the regions of perpetual winter during the glacial period?

The last Indian to dwell on the shores of Lake Hopatcong was Chincopee, an aged warrior of the Nariticongs. Long after the Lenni Lenapé had emigrated from New Jersey this Indian, stirred by memories of his youth, returned to his native waters and hills to spend his last days. He erected a wigwam on what is now Chincopee Cove, and occupied himself fishing and making baskets. But he was finally driven away by the white settlers. With what feelings must the last of the Nariticongs have turned his back upon the lake over whose waters his forefathers had glided in their lithe canoes long before a white man trod upon its shores! Silently he gathered his few belongings into his canoe, and paddled for the last time across the lake to which his race had given its name. Entering the River Styx and penetrating to its furthest recess he vanished in the dark foliage of the mountain forest. The next morning a rift of smoke was seen curling above the trees on the side of the mountain bordering upon the Styx. It was observed at various times for about a week. Then a party was organized to reach the spot and effectually drive away Chincopee from the neighborhood. But on the morning the party was to set out there was no smoke to guide it toward the old Indian's camp. Nevertheless a start was made, and following the trail from the Styx into the woods they came, when half-way up the mountain, suddenly upon the remains of a fire. Near it, bruised, torn and dead, his hand clasping the hilt of his scalping knife, lay Chincopee. Not far from him, stretched on the ground, was the carcass of a huge bear. All about the spot was evidence of the terrible conflict which had proved fatal to both. The wish of the last of the Nariticongs had been fulfilled—he had died on the hunting-grounds of his forefathers. Some superstitious folk believe that his spirit still roams the mountains whose sombre slopes are reflected in the Styx, and sometimes of a morning, when the mists are rising off the lake, they will point to the smoke from Chincopee's camp-fire curling up above the dark foliage.

This picturesque legend of Chincopee is probably a modern adaptation of an older and perhaps equally romantic Indian story—that of Quaquahela, a great sachem, who started to visit a distant tribe of allies in the far South. At sundown he crossed from Pipe Water town to Bonaparte's Landing, and following the shore for a considerable distance, drew up his canoe at what is now known as L'Hommedieu Meadow. From there he set out for the lodge of his friend, Comascoman, who resided on the banks of the Musconetcong, and who was to accompany him on his journey. He had gone but a short distance, when he was attacked by a bear. Bruin being his totem, it was unlawful for him to kill that animal, and he started to return to his canoe ; but the enraged beast prevented this, and a desperate hand to claw battle ensued, in which Quaquahela purchased a victory with his life. A few days later the body of a huge bear was found, and beside it lay the club and totem, and all the hunting gear of the chief. The red men searched for days for their sachem's body, but finding no trace of it, concluded that he had gone to the happy hunting-grounds. The next full moon, the clan

CHINCOPEE COVE.

saw on a side hill a mist ascending to the heavens, and wondered at the strange appearance. That night their medicine man had a vision, and in it Quaquahela appeared and told him that he had erected his spirit-lodge there, and would remain as long as the hills stood. Because of his sin in killing his totem, the bear, he was excluded from the spirit-land forever. He promised to accompany his clan in all their expeditions, and when he retired within his lodge, they would know it by seeing the smoke of his fire ascend to the tops of the trees. He also assured them that every time they would give him a friendly whoop, he would answer. And to this day in damp or wet weather a thin vapor may be seen, rising in curling wreaths over the spot ; and if a shout be given, the answer is heard as distinctly as if Quaquahela himself were replying. Thus the Indians accounted for the echo and vapor which add to the weirdness of the River Styx.

The first white men that settled on Lake Hopatcong were hunters, who are thought to have ventured there about 1775, doubtless tempted by the abundance of game. The woods and mountains abounded with deer, bear and birds, and the lake teemed with fish. One does not have to go back further than 1835 for reminiscences of grand sport in hunting deer at Lake Hopatcong. Sometimes the game was driven into the water and there captured. The severe winter, 1835–36, exterminated the deer, and since then the sport afforded by that region has been limited to quail, partridges and fishing, the black bass of Lake Hopatcong being large and gamy, and there being, besides these, pickerel and perch.

Until within recent years, the charms of Lake Hopatcong were known to only a few of those choice spirits who penetrate into the very solitudes of the forests and mountains in their pursuit of nature's beauties; and this lovely sheet of water lay almost neglected at the threshold of a great city whose population was constantly seeking new fields for summer amusement. The good fishing attracted sportsmen to its shores and it became a favorite camping-out place. But it was not until the Central Railroad of New Jersey purchased and laid out the excursion grounds at Nolan's Point and the Hotel Breslin was built at Chincopee Cove that this lovely sheet of water began to enjoy the measure of popularity it deserved; for, through those enterprises it was able to afford accommodations respectively to people of moderate and ample means; so that now, with the other hotels on the Lake and the opportunities for camping out, Hopatcong attracts people from many and varied walks of life.

The Central Railroad of New Jersey now carries about 50,000 people every summer to Nolan's Point, and there is scarcely a day when the excursion grounds there are not a scene of life and bustle. The amusements afforded at these grounds are many and varied. There are a dancing-pavilion, flying horses and swings; and at the large, commodious float, boats without number. One can go fishing, rowing, sailing or canoeing; or charter a steam-launch and make a tour of the most attractive reaches of the lake. The railroad furnishes the grounds and their appurtenances free of charge, and the fee for the use of boats is low. Parties of half a dozen or more can, for 25 cents each, make the tour of the lake in the steam-launches, and boats can be had for 25 cents an hour. A hot dinner is served at noon at a charge of 50 cents. That Nolan's Point should have become the favorite inland resort for excursions in this section of the country is only the natural result of the beauty of its location, the amusement it affords and the excellence of the railroad service. The company allows no liquor to

be sold on the grounds. It seeks the patronage of respectable people only, and the excursions at Nolan's Point, though they have no end of fun during their day's outing, are notably quiet and orderly. Hence they do not interfere at all with the comfort of private residents or of the guests at the various hotels. Twenty-four hundred acres of water are not easily overcrowded, and so much of an excursion as embarks in boats and launches is soon so scattered over the lake as to lose its identity. Many of the little craft are lost to sight behind the shores of Halsey and Raccoon Islands, in the shadowy recesses of Byram Cove or in the cool shade of the wooded promontories which run, like spurs from the surrounding hills, into the lake. The excursions at Lake Hopatcong, therefore, simply serve to give a joyous zest to the season; the excursionists enjoy themselves so thoroughly—sometimes unmistakably concentrating a whole year's holiday into their one day on the lake—that there is, in some instances, almost a touch of pathos in their unbounded delight.

There is a comfortable hotel at Nolan's Point—the Nolan's Point Villa—which offers excellent accommodations at moderate rates. It is one minute's walk from the railroad station. From the piazza and grounds fine views of the lake are had. Rates of board are $2.00 a day and from $12.00 to $14.00 a week. There are postal and telegraph facilities at Nolan's Point, and a telegraph and telephone office at the "Villa," which also boasts modern sanitary appliances. Before the season opens, the manager, Mr. G. L. Bryant, can be addressed at High Bridge, N. J.; during the season, at the Nolan's Point Villa, Lake Hopatcong, N. J.

The Hotel Breslin gave to Hopatcong its first decided "boom," for it brought to the lake the element of wealth and fashion, in the wake of which everything else follows.

It is a spaciously built house, offering the accommodations of a first-class hotel. Its piazzas are large, and, while strolling up and down them, one can enjoy a lovely view. The lake is reached by a series of terraces, the highest of which is laid out in large flower-beds. To the right, looking from the hotel piazza, is the actress Lotta's cottage, and in the park to the left are many other attractive summer residences. There are the usual amusements of music, dancing, bowling, billiards and tennis. A large boat-house with bath-houses is at the foot of the terraces. Here all kinds of row-boats can be hired, as well as sail-boats and steam-launches. The rates are: Row-boats, per week, $6.00. Day, with man, $3.00; without man, $2.00. Hour, 50 cents; extra hour 25 cents. Sail-boats, per hour, $1.50; extra hour, 50 cents. Fishing tackle and bait can also be had. Steam-launches are let by special contract, the rate for a brief tour being $3.00 an hour. Boats for all trains stop at this point. Connected with the hotel is a livery stable, the rate being $5.00 for a morning or afternoon drive with double team and $4.00 for a single team. A beautiful drive is through Berkshire Valley, of which, by the way, the Jersey Central's passengers obtain a fine view before reaching Minnisink. What is much needed is a road around the lake, which could be made one of the finest drives in the country.

A pretty drive is that to Dover, eight miles distant. Schooley's Mountain, eighteen miles away, is also an interesting drive, taking in Budd's Lake *en route*. It is usually made a day's excursion of.

The rate for transient guests at the Breslin is $5.00 per day ; by the week, $25.00 and $28.00. The headquarters of the management are at the Gilsey House, New York. Post Office address during the season,

which begins about June 15th and continues until October 1, is Rustic, Morris Co., N. J.

According to the map of the State Geologist, Lake Hopatcong is 926 feet above the level of the sea and is the highest navigable lake within 350 miles of New York City. Its shores rise much higher, reaching at one point 1,213 feet. The best view of the lake is to be had from the mountain just above Nolan's Point. Back of Zuck's Lake View House is a look-out, called Friedrich's-Höhe, which would command a superb view if the trees, which now cut off much of the vista, were thinned out. A fair view is also had from Mt. Harry, while from the piazza and upper stories of the Breslin some of the prettiest reaches of the lake can be seen, the hotel being situated about 185 feet above the water.

In taking the reader on a tour of the lake, we will start at the southern end. Where it narrows down toward Shippenport, it is not navigable for steam-launches, and, though the dredging of a channel through this part of Hopatcong has been under consideration for some time, the lake is still entered through a feeder of the Morris Canal. The little launch churns its way gently along between the low banks of the feeder to the lock at Brooklyn, a little settlement which at one time gave the unromantic name of Brooklyn Pond to our lovely Hopatcong. The passage through the lock is an interesting experience. The lower gates are opened and the miniature steamer glides into the lock, the gates closing behind it. It is now imprisoned in a narrow passage. On either side are high, dripping walls, and in front and astern the closed gates. There is a sudden roar of rushing, surging water. The launch lunges half forward, half upward, the churning of the screw adding to the turmoil. The lunging continues, the swashing, surging waters now lifting the launch by the stern, now by the prow. The actions of those who have not been through the lock before are a study. The babies cry; the women grab the nearest man by the arm; the girls are prettily flustered; the men endeavor to appear calm; the passengers that have made the passage before look amused ; the only persons absolutely indifferent are the captain and the engineer—it is such an old story with them that they do not even smile. When the inrushing water has raised the launch eleven feet, almost to a level with the top of the upper gates, these slowly open, and from its narrow, dark and gruesome prison the little vessel glides out upon a lovely reach of water, tranquilly mirroring the wooded slopes of gently rising hills. The change is so sudden that the beholder is almost impelled to rub his eyes, in order to make sure that he is not dreaming or watching the mere shifting of scenes in a theatre. Nature could not devise a more dramatic effect than this sudden emerging from gloom on to the sunlit, dancing waves of Lake Hopatcong.

It is not easy to gain an idea of the length of Lake Hopatcong, for at every point the vista is limited by promontories and islands whose shores overlap one another. For instance, after emerging from the lock, the view up the lake extends, at first, only to Bertrand Island; and it is not until the launch rounds a point on the west shore, a little south of the island, that a passage opens up through which the vessel glides into what seems another and larger lake, for now the reach extends for over two miles to Halsey Island beyond Nolan's Point. About half a mile beyond Bertrand the deep indentation of the Styx opens up on the west and Chincopee and Nepanese Coves on the east; while north of Halsey Island is the reach of three miles to Woodport, and west of it lie Raccoon Island and the broad expanse of Byram and Henderson Coves, the bowl of the original

BERKSHIRE VALLEY ROAD.

"Pipe Water." It is easy to recognize these divisions on the map. They make the lake full of delightful surprises, and, since each new vista differs from the one last enjoyed, while its charms present themselves, as if by a preconcerted plan, in their most attractive order, one becomes at last convinced that Lake Hopatcong is a delicious bit of coquetry on the part of Nature.

As a rule, the hills around the lake are wooded almost to the water's edge so that the shores have a soft and gentle contour. Such is the impression carried away from a general tour of Hopatcong. But the sojourner who has time to make a more minute inspection will discover that the beauty of the shores is varied. One of the most noticeable variations from the velvety softness of the mountain sides is on the west shore, a little beyond Bertrand Island and directly opposite Chincopee Cove. This is Sharp's Rock, which rises boldly about 10 feet out of the water. Its front seems without a crevice for a blade of grass to grow in, and it presents a smooth, solid front, as if engaged only in gazing imperturbably at its own reflection in the lake. It is crowned with a beautiful growth of trees, in whose shade camping-out parties find one of their favorite retreats. The point in which Sharp's Rock is imbedded is known as Tempe Point, a name given to it soon after the Revolution, in honor of Tempe Wick, who lived near Morristown during the winter of 1779, when Washington's army was quartered among the Morris County hills. Tempe was a noted horsewoman—a skillful and daring rider. At a time when the Americans were securing all the horses they could for the army, a troop of horse came in sight of her as she was riding her favorite steed. Realizing what would be the animal's fate if the troop came up with her, she turned, and putting the whip to her horse, dashed away at full speed toward home, pursued by the soldiers. She so distanced them that she had time, before they reached the house and began searching the stable, to lead the horse upstairs and secrete it in an apartment, where she kept it for several weeks, until all danger of its being seized had passed.

Right around the point from Sharp's Rock is the entrance to the Styx. This is one of the weirdest and most romantic retreats of the lake. For about half a mile from its entrance, one cannot comprehend why it should have been named after the river of Hades. But beyond the point where it is crossed by a bridge, it divides itself into two arms, running north and south. Each of these, gradually narrowing so that the foliage on the hills casts its shadow from shore to shore and the mountains seem to close in upon the sullen waters, loses itself in the gloomy recesses of the forest. Trunks of decaying trees, whose jagged branches protrude here and there above the surface of the black waters ; the stillness, broken only by the shrill cry of the kingfisher and its unearthly echoes among the hills ; the long, hairy grasses which, as the water is disturbed by the splash of the oars, rise from the bottom, trail after the boat a moment, and then vanish with sinuous, snake-like movements, combine to so impress the traveler on the Styx with a sense of the mysterious and supernatural that he is ready to accept, without questioning, the legend of the Nariticong whose spirit is said to haunt the depths of the forest back of the south arm of the inlet.

Chincopee Cove, on the opposite side of the lake, is a pretty nook. Its north shore is formed by Chestnut Point, a low, thicky-wooded promontory, tapering off to a narrow tongue covered with a beautiful grove, between whose branches one sees glints of sunlit ripples beyond. Across the lake is the gray tablet of Sharp's Rock and the sombre shores of the entrance to the Styx. To the south this love-

CANAL FEEDER.

ly view is rounded off by Bertrand Island. In fact, Chincopee Cove is another of those little lake vistas to which Hopatcong owes so much of its charm. Not until the oarsman has pulled out beyond Chestnut Point and Bertrand does he realize that he has been on a lake within a lake. On this cove, so retired, yet so near the main reach, the Hotel Breslin is situated ; so that, added to the picturesque features of Chincopee, are the life on the beautiful grounds which surround the hotel and the dashes of color made by the brilliant costumes of the young people (for, thanks to blazers, sashes and tennis-shirts, it is now with the genus homo as with other creatures of nature--the male can be as gaudy as the female, or even gaudier).

Adjoining Chincopee Cove and lying between Chestnut and Nolan's Points, is Nepanese or Big Cove. The vicinity of Nolan's Point is well settled with summer residences. On the point itself, besides the excursion grounds and hotel already mentioned, and the railroad buildings, are ice houses of great capacity, the cutting and shipping of ice from the pure frozen waters of Lake Hopatcong having developed into an industry of such proportions that winter only varies the operations of the railroad. The Central Railroad Company of New Jersey's branch from High Bridge to Lake Hopatcong is by no means a spur for summer traffic only. The long excursion trains of summer are succeeded by the longer ice trains of winter. This branch derives additional importance from its connection at Nolan's Point with the Ogden Mine Railroad, now also part of the Jersey Central's system. This road extends about nine miles north of the lake to the Ogden Mine, with stations at the famous Hurd Mine, whose sloping shaft runs 3,800 feet into the mountain, and to a depth of about 1,800 feet in a direct line, and at the Weldon and Ford and Schofield

Mines. From an elevation, but a few steps from the station at the Ogden Mine, a superb view is had in the direction of the Delaware Water Gap, which is well worth a trip from the lake. Another pretty trip for sojourners at Lake Hopatcong, is to take the Ogden Mine Railroad to Hopewell Crossing, about two and a quarter miles before the terminus at Ogden is reached, and where the train will be stopped at the request of any passenger who wishes to alight there. Taking the road to the right of the railroad, one reaches, after walking about one and one-eighth of a mile, Morris Pond, a lovely little sheet of water, noted for its crystalline clearness and the beauty of its shores, the background of the view from the south end being a high mountain, whose base is laved by the waters of this charming lake. Picnic parties will find a pretty little island attractive. There are many fine black bass in Morris Pond, but, owing to the clearness of its waters, fishing is good only on a cloudy day.

Off Nolan's Point, a little northward, is the emerald circle of Halsey Island. Thickly covered with hemlock, spruce and pine, it looks like a soft tuft of green floating on the water. Some of the spruces on this island, which is, like Sharp's Rock, a favorite camping-ground, are three feet in diameter. Halsey Island lies at the entrance of a large bay, which embraces Byram and Henderson Coves with their nooks and corners among pretty islets and rocky ledges. Passing through the little sound formed by Halsey and Raccoon Islands with the southerly shore of this bay, one sees, on the main, Bishop's Rock, a long, picturesque mass, near which is the boulder where Bonnel Moody, a notorious leader of England's partisans during the Revolution, concealed himself while Brandt, the Mohawk Chief, visited the Indian village opposite and induced the Nariticongs to take part in the massacre of the Minnisink. The point

of land just beyond Bishop's Rock to the northwest, rising some thirty feet and crowned by a pine, tall and vigorous in its old age, is known as Bonaparte's Landing. Joseph Bonaparte, ex-King of Naples and Spain and the brother of Napoleon Bonaparte, landed here while on a prospecting tour, which also included Budd's Lake. (See p. 86.)

For wild, rugged beauty Byram and Henderson Coves surpass the other reaches of the lake. The shores rise from one to two hundred feet, and terraces of rock have rent and torn the green mantle which the forest has flung over the mountains. Most striking are the ledges which form the Devil's Stairs, to which Brandt and Moody paddled after the former's return from the Indian village. On the southern shore of Byram Cove is Sperry Spring, a rill of clear, icy water. Henderson Rock is a great boulder on the point of the eastern shore of Henderson Cove, of which Raccoon Island seems a continuation, the passage between being so narrow. This island is wooded like Halsey, but its shores project here and there in picturesque ledges. The shores of the narrow northern arm of the lake, which runs to Woodport, are low and fringed with meadows and woodland. Their delicate beauty is another of Hopatcong's varied charms.

In the shallow little arm which runs at the southerly end of the lake, parallel with the canal feeder, is Floating Island, a natural curiosity which attracts considerable attention from those interested in such matters. This island, whether at low or high water, always remains about one foot above the level of the lake. It is noted for its flora, which includes the "side-saddle plant" (Sarracenia) with its pitcher-like leaves, cornplants, sweet-briar roses, the rhododendron, spruce and tamarack. The Musconetcong River, of whose picturesque valley we have had so many glimpses from Schooley's Mountain, is the outlet of the lake.

The camps add not a little to the charms of life at Hopatcong. The white tents gleaming among the trees, the fluttering flags and pennants, the boats moored at the landings, anchored at the fishing grounds or speeding along to the splash and cadence of the oars, the sun-browned faces of the happy campers, combine to make those who are leading a less Bohemian existence enter with greater zest into the enjoyment of their outing. The camps play an important part in the annual Harvest Moon Festival, which is celebrated by an illumination. Colored lanterns glimmer among the trees and on the boats ; and nearly every camp has its display of fire-works. An illuminated boat parade is another feature. As the boats are invisible, the colored lights are like will-o'-the-wisps, of all hues, floating in and out of coves, among islands and up and down the lake, while voices of unseen singers are borne over the water. As the night wears on the lights separate and glide in many directions toward the various camps, cottages and hotels; one by one they vanish in the dark shadow of the shores, and voice after voice grows silent. Then, when the lake lies like a mirror in the moonlight, there is heard a murmuring among the hills, like a low, melodious chanting of many distant voices; and one half fancies that the nymphs and naiads have come forth from the shaded springs and rills of the forest and gathered upon some moonlit meadow far up on the mountain to celebrate the glories of the night.

INDEX.

Spotted something that isn't quite right?

Let us know and we'll fix it for future reprints. And as a thank you for your diligence, we'll send you a free Sothis book of your choice.

publisher@sothis.press

This page intentionally left blank

Made in the USA
Coppell, TX
05 November 2026

11292665R00132